Soil Science and Agriculture : Theory and Practice

Soil Science and Agriculture Theory and Practice

Dr. N.K. Rajan

AMIGA PRESS. INC
Delhi -110051

AMIGA PRESS. INC

H.O :- 28-A, Indra Park, Chander Nagar

Delhi :- 110051 (India)

Tel :- 011-22504028, +91-98736-57000

Email :- amigapress@gmail.com

amigaeditorial@gmail.com

amigapub@gmail.com

Website :- www.amigapress@co.in

This edition is published by Amiga Press.Inc

ISBN : 978-93-84533-34-2

First Published 2018

PRINTED : THOMSON PRESS INDIA

Preface

Soil Science is an important branch of agriculture which deals with studies of different areas like soil physics, soil chemistry, soil survey, genesis and classification, soil mineralogy, etc. It is, therefore, essential that this valuable natural resource is understood and conserved properly for sustained agricultural production in India. National agricultural production is mainly dependent upon one of the important natural resource like Soil the maintenance and improvement of soil productivity for sustained agricultural production necessitates acquisition though knowledge of oil science.

Agricultural soil science follows the holistic method. Soil is investigated in relation to and as integral part of terrestrial ecosystems but is also recognized as a manageable natural resource. Agricultural soil science studies the chemical, physical, biological, and mineralogical composition of soils as they relate to agriculture. Agricultural soil scientists develop methods that will improve the use of soil and increase the production of food and fiber crops. Emphasis continues to grow on the importance of soil sustainability.

Agricultural soil scientists study ways to make soils more productive. They classify soils and test them to determine whether they contain nutrients vital to plant growth. Such nutritional substances include compounds of nitrogen, phosphorus, and potassium. If a certain soil is deficient in these substances, fertilizers may provide them. Agricultural soil scientists investigate the movement of nutrients through the soil, and the amount of nutrients absorbed by a plant's roots. Agricultural soil scientists also examine the development of roots and their relation to the soil. Some agricultural soil scientists try to understand the structure and function of soils in relation to soil fertility.

In addition, agricultural soil scientists develop methods to preserve the agricultural productivity of soil and to decrease the effects on productivity of erosion by wind and water. For example, a technique called contour plowing may be used to prevent soil erosion and conserve rainfall. Researchers in agricultural soil science also seek ways to use the soil more effectively in addressing associated challenges. Such challenges include the beneficial reuse of human and animal wastes using agricultural crops; agricultural soil management aspects of preventing water pollution and the build-up in agricultural soil of chemical pesticides.

Although agriculture is essential for human food and the stability of complex societies, almost all of our evolution has taken place in small, mobile, kin-based social groups, such as bands and tribes. Before we became sedentary people dependent on agriculture, we were largely dependent on wild plant and animal foods, without managing soil and water resources for food production. Our social evolution has accelerated since the Agricultural Revolution and taken place synergistically with human biological evolution, as we have become dependent on domesticated plants and animals grown purposefully in highly managed, soil-water systems.

This book primarily provides comprehensive overview on the role and behaviour of soils in both the man-made and the natural environment. Many farmers of India and abroad are not fully conversant with the methods of keeping good soil health and improving soil-fertility. Because of harmful side effect of chemical fertilizers we must also give encouragement to the farmers to use organic manures in the agricultural field without application of pesticidal compounds.

—*Editor*

Contents

1

Soil Science

Soil science is the study of soil as a natural resource on the surface of the earth including soil formation, classification and mapping; physical, chemical, biological, and fertility properties of soils; and these properties in relation to the use and management of soils. Sometimes terms which refer to branches of soil science, such as pedology (formation, chemistry, morphology and classification of soil) and edaphology (influence of soil on organisms, especially plants), are used as if synonymous with soil science. The diversity of names associated with this discipline is related to the various associations concerned. Indeed, engineers, agronomists, chemists, geologists, geographers, ecologists, biologists, microbiologists, sylviculturists, sanitarians, archaeologists, and specialists in regional planning, all contribute to further knowledge of soils and the advancement of the soil sciences. Soil scientists have raised concerns about how to to preserve and soil and arable land in a world with a growing population, possible future water crisis, increasing per capita food consumption, and land degradation.

FIELDS OF STUDY

Soil occupies the pedosphere, one of Earth's spheres that the geosciences use to organize the Earth conceptually. This is the conceptual perspective of pedology and edaphology, the two main branches of soil science. Pedology is the study of soil in its natural setting. Edaphology is the study of soil in relation to soil-dependent uses. Both branches apply a combination of soil physics, soil chemistry, and soil biology. Due to the numerous interactions between the biosphere, atmosphere and hydrosphere that are hosted within the pedosphere, more integrated, less soil-centric concepts are also valuable. Many concepts essential to understanding soil come from individuals not identifiable strictly as soil scientists. This highlights the interdisciplinary nature of soil concepts.

Research

Dependence on and curiosity about soil, exploring the diversity and dynamic of this resource continues to yield fresh discoveries and insights. New avenues of soil research are compelled by a need to understand soil in the context of climate change, greenhouse gases, and carbon sequestration. Interest in maintaining the planet's biodiversity and in exploring past cultures has also stimulated renewed interest in achieving a more refined understanding of soil.

Mapping

Most empirical knowledge of soil in nature comes from soil survey efforts. Soil survey, or soil mapping, is the process of determining the soil types or other properties of the soil cover over a landscape, and mapping them for others to understand and use. It relies heavily on distinguishing the individual influences of the five classic soil forming factors. This effort draws upon geomorphology, physical geography, and analysis of vegetation and land-use patterns. Primary data for the soil survey are acquired by field sampling and supported by remote sensing.

Classification

As of 2006, the World Reference Base for Soil Resources, via its Land & Water Development division, is the pre-eminent soil classification system. It replaces the previous FAO soil classification. The WRB borrows from modern soil classification concepts, including USDA soil taxonomy. The classification is based mainly on soil morphology as an expression pedogenesis. A major difference with USDA soil taxonomy is that soil climate is not part of the system, except insofar as climate influences soil profile characteristics. Many other classification schemes exist, including vernacular systems. The structure in vernacular systems are either nominal, giving unique names to soils or landscapes, or descriptive, naming soils by their characteristics such as red, hot, fat, or sandy. Soils are distinguished by obvious characteristics, such as physical appearance (e.g., color, texture, landscape position), performance (e.g., production capability, flooding), and accompanying vegetation. A vernacular distinction familiar to many is classifying texture as heavy or light. Light soil content and better structure, take less effort to turn and cultivate. Contrary to popular belief, light soils do not weigh less than heavy soils on an air dry basis nor do they have more porosity.

Vasily Dokuchaev, a Russian geologist, geographer and early soil scientist, is credited with identifying soil as a resource whose distinctness and complexity deserved to be separated conceptually from geology and crop production and treated as a whole. Previously, soil had been considered a product of chemical transformations of rocks, a dead substrate from which plants derive nutritious elements.

Soil and bedrock were in fact equated. Dokuchaev considers the soil as a natural body having its own genesis and its own history of development, a body with complex and multiform processes taking place within it. The soil is considered as different from bedrock. The latter becomes soil under the influence of a series of soil-formation factors (climate, vegetation, country, relief and age). According to him, soil should be called the "daily" or outward horizons of rocks regardless of the type; they are changed naturally by the common effect of water, air and various kinds of living and dead organisms. A 1914 encyclopedic definition: "the different forms of earth on the surface of the rocks, formed by the breaking down or weathering of rocks". serves to illustrate the historic view of soil which persisted from the 19th century. Dokuchaev's late 19th century soil concept developed in the 20th century to one of soil as earthy material that has been altered by living processes. A corollary concept is that soil without a living component is simply a part of earth's outer layer. Further refinement of the soil concept is occurring in view of an appreciation of energy transport and transformation within soil. The term is popularly applied to the material on the surface of the Earth's moon and Mars, a usage acceptable within a portion of the scientific community. Accurate to this modern understanding of soil is Nikiforoff's 1959 definition of soil as the "excited skin of the sub aerial part of the earth's crust".

BIODYNAMIC

Seed Soaks and Sprays

Biodynamic preparations are used to enhance the biological activity of the soil. The preparations consist of mineral, plant, or animal manure extracts, usually fermented and applied in small proportions to compost, manures, the soil, or directly onto plants, after dilution and stirring procedures called dynamizations. They are numbered 500 to 508. BD#500 is made from cow manure fermented in a bovine horn and buried for six months through autumn and winter. It is used as a soil spray to stimulate root growth and the production of humus. BD#501 is the horn silica preparation made from powdered quartz packed into a cow horn and buried in the soil through spring and summer for six months. It is used as a plant spray for the stimulation and regulation of growth. BD#502 - 507 are compost inoculates and are inserted into the compost piles to increase microbial activity, enhancing the decomposition process. They are made from the fermented herbs yarrow (*Achillea millefolium*), chamomile(*Anthemis nobilis*), stinging nettle(*Urtica dioica*), oak bark (*Quercus alba*), dandelion (*Taraxacum officinale*), and valerian (*Valariana officinalis*), respectively. Each preparation stimulates processes essential for plant growth and are used to strengthen the life forces on the farm. The preparations are used a seed soak to encourage germination and growth of seedlings and for anti-fungal control. The solutions are lightly sprayed on seeds and then quick dried on a screen.

Recommended by Gunther Hauk:

BD 501, horn silica prep: Stir for I hour. Spray on seed crop a fine mist early in the morning on a fruit day. The week before harvest, stir and spray finely in the afternoon. Refer to BD Planting Calendar.

BD 500 - applied to soil before the sowing day. The vegetable seeds are treated with BD 500 for about 12 an hour and dried in the shade for about 1 hour. The seeds are then sown in the beds. Healthy plants are obtained and the germination of seeds is very good. Seeds treated with this method include Cowpea, Okra, Radish, Beans.

Herbal Treatments

Equisitum, common name: Field Horsetail, is prepared as a tea to suppress fungal growth of mildew and other fungi, on grapevine, vegetables and trees. The useful variety, 'arvense' has a slightly discolored or green collar. The less potent varieties have a collar formed of small black pointed leaves around the joints. The plant propagates by spores that spread out from the brown leafless stems in spring. The plant is difficult to start but reproduces well once established. Prepare by boiling plants for a 20 minutes to a half an hour. Spray on seeds.

Natural Products

'Considerable research activity has occurred in the Asian-Pacific region on the potential for plant extracts to control seed-borne fungi. The oils of cassia and clove inhibited growth of established seed borne infections *of Aspergillus flavus, Curvularia pallescens* and *Chaetomium indicum* in maize (Chatterjee 1990). Aqueous extracts of *Strychnos nux-vomica,* garlic bulbs, ginger rhizomes, basil leaves, and fruits of *Azadirachta indiica* were used to control *Alternaria padwickii* in rice seeds (Shetty et al 1989), while extracts from peppermint and garlic reduced rice seed infection by *Cochliobolus miyabeanus* (Alice and Rao 1986). Garlic bulb extract inhibited the spore germination and mycelial growth of seed-borne fungal pathogens of jute, including *Macrrophomina phaseolina, Botryodiplodia theobromae*and *Colletotrichum corchori* (Ahmed and Sultana 1984). Homeopathicdrugs, *Filixmas*and *Blatta orientalis,* completely suppressed the population of *Fusariumoxysporum* in the seed mycoflora of wheat (Raka et al. 1989). *Aspergillus ruber* infection and weevil oviposition of *Zabrotes subfaciatus* were reduced by mineral oil and soybean oil treatment of dry beans stored in Ecuaddor (Hall and Harman 1991). Soybean oil, applied at a rate used to suppress grain dust, reduced storage fungi growth in maize and soybeans during 12 months in field storage bins in Iowa (McGee et al 1989, White and Toman 1994).'

Hot Water Seed Treatment

kills most disease causing organisms on or within seed. Suggested for eggplant, pepper, tomato, cucumber, carrot, spinach, lettuce, celery, cabbage, turnip, radish, and other crucifers

- Use carefully: Improper treatment can cause seed injury
- Seed of cucurbits can be severely damaged by hot-water treatment

Pre-warm loose seed in woven cotton bag (about _ full) for 10 minutes at 100 F water. Place pre-warmed seed in water bath that will hold the recommended temperature.

Length of treatment must be EXACT.

After treatment, dip bags in cold water to stop heating action, spread seeds to dry. Recommends applying protective seed tx fungicide to hot-water treated seed old seed can be severely damaged by this treatment. A small sample of any seed lot over 1 year old should be treated and tested for germination to determine amount of injury that may occur. Thiram is most frequently suggested seed-protectant fungicide.

Hot Water Treatment of Vegetable Seeds

The following is an excerpt from Sally A. Miller's "Disease Management for Conventional and Organic Tomato Growers", from the 2002 New York State Vegetable Conference and Berry Growers Meeting Proceedings.

"*Cultural management.* This is a broad category of management tactics that are numerous common-sense approaches that affect not only disease incidence and severity, but also plant growth and management of other pests, in both the greenhouse and field. The first consideration is to always use clean seeds, as free of pathogens as possible. For tomatoes, seeds should be tested for the bacterial pathogens causing bacterial canker, bacterial spot and bacterial speck.

Certain fungi may also be seed borne, but testing is not routinely carried out for these pathogens. We recommend hot water treatment for all seed lots testing positive for a bacterial pathogen, and all untested seed lots. Hot water treatment is preferred because the bacterial canker pathogen survives inside the seed coat and is not completely eliminated by surface disinfestations using Clorox, acid, or other treatments. It is also permitted for organic tomato production. Tomato seeds should be placed in a loosely woven cotton (e.g. cheesecloth) bag, not more than half-full, and pre-warmed for 10 minutes in 100 degree Fahrenheit water. The bag should then be moved to a 122 degree Fahrenheit water bath for 25 minutes exactly. Old or poor quality seed may be damaged by this treatment. For more complete instructions.

Preventing Bacterial Diseases of Vegetables with Hot-Water Seed Treatment

Hot Water Treatment

Hot water treatments control many seed-borne diseases by using temperatures hot enough to kill the organism but not quite hot enough to kill the seed. Treatment for the fungal disease blackleg and the bacterial disease black rot of crucifers is a classic example of hot water treatment. It must be carefully and accurately done. Therefore, growers should ensure that heat treatment was performed by the seed company, avoiding performing hot water treatments themselves. A few degrees cooler or hotter than recommended may not control the disease or may kill the seed. The objective is to use as high a temperature and as long a duration as possible without seriously impacting germination. Hot water treatment can be damaging or not practical for seeds of peas, beans, cucumbers, lettuce, sweet corn, beets and some other crops. Some hybrid varieties of cauliflower may be damaged by the recommended treatment. Seeds that can be treated by hot water are listed in the table.

Methodology For General Reference

Hot water treatment of fresh seed at the temperatures recommended should not reduce germination. However, check carefully with suppliers to ensure that the seed has not already been treated by the seed company. Seed should not be treated twice. Also, treating old, out-of-date seed will reduce germination. These steps must be followed for accurate treatment:

- Place a few grams of seed in a small porous bag, such as cheesecloth. The amount of seed should be just sufficient to allow thorough and immediate wetting. The bag may need to be weighted down.

- Fill a container with water and increase the temperature. Use an accurate thermometer to check the temperature and immerse the thermometer to half way down the container. It is absolutely critical that time and temperature be monitored precisely as high temperatures can kill seeds while low temperatures do not kill the pathogen.
- When the water reaches the correct temperature, wet the bag and seed with warm water and suspend them in the container of water.
- Stir the water and the bag of seed regularly during treatment to ensure that all the seed is heated evenly.
- After the time/temperatures have been satisfied, immediately place the seeds (bag & all) into cold water to quickly stop the heating.
- Once seeds have cooled, spread them thinly on a paper towel to allow drying.
- Always use a protective seed treatment fungicide with hot-water-treated seeds after they dry, for these seeds have been weakened and are more susceptible to soil-borne pathogens.
- Plant the seed as soon as it is thoroughly dry. Do not store treated seed.

Spores or bacteria that are attached to seeds can be killed by soaking the seeds in hot water. Use water of exactly 50°Celcius and soak the seeds for 30 minutes.

Disinfectants

Clorox Treatment: removes bacterial pathogens on seed surface. Recommended for peppers, tomatoes, cucurbits and other vegetavles if the seed has not been treated by another method To use for organic seed, this is NOT a treatment, but seed cleaning. Rinse with water after the washing with bleach. Ask your certifier. Instructions: Agitate seeds in solution of 1 part Clorox in 4 parts water with one teaspoon surfactant for 1 minute. Use 1 gallon disinfectant solution per pound of seed (prepare a new solution with each batch) Rinse seed thoroughly in running tab water for 5 minutes, then spread and dry. Dust seeds with Thiram (75 WP 1 teaspoon/lb seed). Do treatment close to planting time.

Seed Soaked in Bleach or Acid

Soaking seed in a bleach or acid solution can be very helpful in control of certain bacterial pathogens that may be carried on the seed coat, including bacterial speck, canker, and spot or tomato or pepper. This is a treatment that growers can and should implement themselves. The time and concentration needed varies by crop.

Clorox Seed Treatment for Tomato and Pepper

Seed may be treated by washing 40 mins. w/continuous agitation in 1 part Clorox liquid bleach (5.25% sodium hypochlorite) to 4 parts water (i.e. 1 pint Clorox plus 4 pints water). Rinse seed in clean water immediately after removal from the Clorox solution and promptly dry. Germination may be compromised if washing time exceeds 40 mins.

Commercial Products

Kodiak concentrate contains Bacillus subtilis bacteria which colonize the developing root system, suppressing disease organisms such as Fusarium, Rhizoctonia, Alternaria and Aspergillus that attack root systems. When used with a chemical seed treatment, the combination of chemicals and Kodiak provides protection to the root for a much longer time than with chemicals alone. As the root system develops, the bacteria grow with the roots extending the protection throughout the growing season. As a result of this biological protection, a vigorous root system is established by the plant, which often results in more uniform stands and greater yields. Registered for seed and pod vegtables, soybeans, wheat and barley, and corn plus all other agricultural seeds.

Biological Seed Treatment for Corn and Soybeans

Bio-Seed-Gard is an OMRI-approved blend of micro-organisms for use as a seed treatment for Corn, Soybeans, Peas, Legumes, and other crops. AgriEnergy Resources, the Co. that developed the product, has done greenhouse and field trials showing improved: seeding vigor, stand establishment, root growth, plant growth, and yield. Dry blend of Trichoderma & Mycorrhizal species. CB-QGG is a liquid biological seed treatment and root growth promoter formulated with beneficial microbes, macro and micro nutrients, amino acids, organic acids, root growth stimulants, enzymes, proteins, vitamins and minerals. CB-QGG makes less available forms of soil phosphate available to plants, promotes nitrogen fixation, root development and quick emergence, stimulates cell division and increases stress tolerance. In addition, it produces substances which increase the vigor of treated plants helping them resist damage by pathogenic fungi and damping off. This results in a more consistent plant stand and increased yields. CB-QGG promotes early root growth and larger roots. Larger roots provide better access to moisture and nutrients which translates into improved health throughout the life cycle of the crop. When used as directed, the beneficial fungi in CB-QGG penetrate and colonize the plant roots and send out filaments into the surrounding soil. These filaments form a bridge that connects the plant roots with large areas of soil (up to 200 times larger than the root zone) and act as a "pipeline" to funnel nutrients to the plant. In return, the plant discharges compounds, through its roots, to stimulate fungal growth. The efficacy of the fungi is also enhanced by soil microbes such as bacillus sp. These bacteria, many of which are aerobic, affect root colonization and function by producing vitamins, hormones, and other compounds that promote fungal growth.

Biological Seed Treatment

Fungal spores or bacteria can infect seeds while they are still developing on the plant or after the harvest. These pathogens cause diseases in the next crop (seed-borne diseases). Antagonistic fungi or bacteria can be used to protect seeds. Examples of these biological agents are *Trichoderma* sp. (antagonist fungus) and *Bacillus subtilis* (a bacterium). Biological seed treatments multiply in the soil, protecting root systems against soil-borne pathogens (ie: damping-off) after germination. Trace Element SeedTreatment 'Application of trace elements to the seed provides nutrients to the seed during the critical period of plant establishment. The early alleviation of nutrient deficiency enhances crop yield and quality.'

Indigenous

Seed Treatment Methods

Seed Treatment 1 : The cow urine treatment with 1:10 concentration was found very suitable to treat seeds of finger millet for good germination and seedling vigor. So farmers were advised accordingly to treat the finger millet with cow urine at 1:10 dilution.

Seed Treatment 2 : Finger millet seeds are soaked with water for 18 hours (1 kg seed/600 ml water). The water is drained and seeds are dried under shade just 10 - 15 min before sowing. It helps in good germination and through this treatment it resists stress and overcomes drought condition very effectively.

Seed Treatment 3 : Salt Water Treatment (Place an egg in water until it sinks to the bottom and rests horizontally. The horizontal egg indicates that the egg is not rot and that it could be further used for purpose of testing the density of salt concentration in water). Then remove the egg add salt to water and stir continuously until it dissolves. Immerse the egg once again into this concentration. This process of immersing the egg to test the density of salinity concentration continues. At a certain stage of salt application into water, only a quarter of the egg is found to be visible on the surface. This is the standard saline concentration used for seed treatment. Seeds are soaked in this preparation for a period of 30 minutes and then removed, washed twice and thoroughly dried until the moisture is absorbed. Salt-water treatment is a useful technique in separating healthy seeds from chaffy ones. This treatment is found to be very effective in treating paddy seeds.

1. It separates the healthy seeds from chaffy seeds
2. Protects from storage fungus
3. Activates embryo

Biological Seed Treatment Research

Development of Biological Seed Treatments

A European Commission funded project involving Aberdeen University (Dr Chris Mullins), TCOA (Prof. Carlo Leifert), a German seed company and Universities in Germany and Greece is close to completion. Over the last 3 years processing methods, which allow biological control agents and plant extracts to be applied as a seed coating. This novel technology replace fungicide and pesticide seed treatments used in conventional agriculture to protect against soil borne diseases.

AREAS OF PRACTICE

Academically, soil scientists tend to be drawn to one of five areas of specialization: microbiology, pedology, edaphology, physics or chemistry. Yet the work specifics are very much dictated by the challenges facing our civilization's desire to sustain the land that supports it, and the distinctions between the sub-disciplines of soil science often blur in the process. Soil science professionals commonly stay current in soil chemistry, soil physics, soil microbiology, pedology, and applied soil science in related disciplines

One interesting effort drawing in soil scientists in the USA as of 2004 is the Soil Quality Initiative. Central to the Soil Quality Initiative is developing indices of soil health and then monitoring them in a way that gives us long term (decade-to-decade) feedback on our performance as stewards of the planet. The effort includes understanding the functions of soil microbiotic crusts and exploring the potential to sequester atmospheric carbon in soil organic matter. The concept of soil quality, however, has not been without its share of controversy and criticism, including critiques by Nobel Laureate Norman Borlaug and World Food Prize Winner Pedro Sanchez.

A more traditional role for soil scientists has been to map soils. Most every area in the United States now has a published soil survey, which includes interpretive tables as to how soil properties support or limit activities and uses. An internationally accepted soil taxonomy allows uniform communication of soil characteristics and functions. National and international soil survey efforts have given the profession unique insights into landscape scale functions. The landscape functions that soil scientists are called upon to address in the field seem to fall roughly into six areas:

- Land-based treatment of wastes
  - o Septic system
  - o Manure
  - o Municipal biosolids
  - o Food and fiber processing waste
- Identification and protection of environmentally critical areas
  - o Sensitive and unstable soils
  - o Wetlands
  - o Unique soil situations that support valuable habitat, and ecosystem diversity
- Management for optimum land productivity
  - o Silviculture
  - o Agronomy
    - Nutrient management
    - Water management
  - o Native vegetation
  - o Grazing
- Management for optimum water quality
  - o Stormwater management
  - o Sediment and erosion control
- Remediation and restoration of damaged lands
  - o Mine reclamation
  - o Flood and storm damage
  - o Contamination
- Sustainability of desired uses
  - o Soil conservation

There are also practical applications of soil science that might not be apparent from looking at a published soil survey.

- Radiometric dating: specifically a knowledge of local pedology is used to date prior activity at the site
    - o Stratification (archeology) where soil formation processes and preservative qualities can inform the study of archaeological sites
    - o Geological phenomena
    - - Landslides
    - - Active faults
- Altering soils to achieve new uses
    - o Vitrification to contain radioactive wastes
    - o Enhancing soil microbial capabilities in degrading contaminants (bioremediation).
    - o Carbon sequestration
    - o Environmental soil science
- Pedology
    - o Soil genesis
    - o Pedometrics
    - o Soil morphology
    - - Soil micromorphology
    - o Soil classification
    - - USDA soil taxonomy
- Soil biology
    - o Soil microbiology
- Soil chemistry
    - o Soil biochemistry
    - o Soil mineralogy
- Soil physics
    - o Pedotransfer function
    - o Soil mechanics and engineering
- Soil hydrology, hydropedology

Fields of Application in soil science

- Soil survey
- Soil management
- Standard methods of analysis
- Soil fertility / Nutrient management
- Ecosystem studies

- Climate change
- Watershed and wetland studies
- Pedotransfer function

Related disciplines

- Agricultural sciences
  - o Agrophysics science
  - o Irrigation management
- Anthropology
  - o archaeological stratigraphy
- Environmental science
  - o Landscape ecology
- Geography
  - o Physical geography
- Geology
  - o Biogeochemistry
  - o Geomicrobiology
  - o Geomorphology
- Hydrology
  - o Hydrogeology
- Waste management
- Wetland science

Agrology

Agrology is the branch of soil science dealing with the production of crops. The use of the term is most active in Canada. Use of the term outside of Canada is sporadic but significant. The term appears especially well established in Russia and China, with agrologists on university faculty lists and agrology curricula. Agrology is synonymous with agricultural science when used in Canada, is nearly synonymous with the U.S. term *agronomy*, and has a meaning related to soil science when used outside of Canada.

Canada

The term agrologist was coined by Dr. J. B. Harrington and adopted in 1946 to fill the need in Canada to have a term to denote "provincial agriculturalist". The title of Professional Agrologist is conferred on persons with at least a Bachelor's Degree in Agriculture and who can demonstrate the qualities needed to responsibly teach, practise, or conduct experiments and research in the agricultural sciences. According the Agricultural Institute of Canada website, an agrologist can also hold a degree in a field related to agriculture, or in some provinces pass rigorous prescribed examinations to attain a professional designation.

There are about 5000 agrologists in Canada as of 2004. Within British Columbia the term "agrology" is defined by an Act of the Legislature passed in 2003 and adopted in 2004 entitled the Agrologists Act. This Act authorizes the self-governing body, the British Columbia Institute of Agrologists and those practising agrology within British Columbia do so under the following definition. "Agrology" means using agricultural and natural sciences and agricultural and resource economics, including collecting or analyzing data or carrying out research or assessments, to design, evaluate, advise on, direct or otherwise provide professional support to

(a) the cultivation, production, improvement, processing or marketing of aquatic or terrestrial plants or animals, or

(b) the classification, management, use, conservation, protection, restoration, reclamation or enhancement of aquatic or terrestrial ecosystems that are affected by, sustain, or have the potential to sustain the cultivation or production of aquatic or terrestrial plants or animals;

BCIA, the governing body for Agrologists in British Columbia, has over 1000 registered members. The Registrars of Professional Agrologists across Canada adopted the following definition of Agrology in May 2007. Agrology is the practice of bioresource sciences to provide knowledge and advice to support the development of the agriculture sector and the health of the society, environment, and economy.

Outside Canada

Outside of Canada, the term agrology is synonymous with soil science and is not in common usage in English-speaking countries.

Agrology in Soil Science Society Glossaries

Two national member societies (Canadian, American) of the International Union of Soil Sciences (IUSS) maintain and publish glossaries of scientific terms. Other soil science societies defer to the American glossary. The term agrology is not in use. Edaphology or crop edaphology in combination with soil management would be the preferred approach used by soil scientists to concisely describe soil science as it applies to crop production.

Agrology dictionary definitions

As of 2004, no dictionary definition of agrology is yet consistent with the Canadian use of the term and dictionary definitions fall into one of four categories.

1. agrology is defined as synonymous with soil science. The root agr- is represented as meaning soil.
2. agrology is defined as synonymous with soil science, but the context implies that soil science is a subdiscipline of agricultural science.
3. agrology is defined as the subdiscipline of soil science as it applies to crop production. This would make agrology synonymous with the term crop edaphology.
4. agrology is defined as the subdiscipline of agronomy that considers the influence of soil.

Agricultural Land

Agricultural land (also agricultural area) denotes the land suitable for agricultural production, both crops and livestock. It is one of the main resources in agriculture. The standard classification (used, e.g., by FAO — Food and Agriculture Organization of the United Nations) divides agricultural land into the following components:

- Arable land — land under annual crops, such as cereals, cotton, other technical crops, potatoes, vegetables, and melons; also includes land left temporarily fallow.
- Orchards and vineyards — land under permanent crops (e.g., fruit plantations).
- Meadows and pastures — areas for natural grasses and grazing of livestock.

The first two components — arable land and land in permanent crops — constitute so-called *cultivable land*. The part of arable land actually under crops is called *sown land* or *cropped land*. The term *farmland* is ambiguous in the sense that it may refer, on the one hand, to agricultural land and, on the other hand, to cultivable or even only arable land. Depending on the use of artificial irrigation, agricultural land is divided into irrigated and non-irrigated land. In arid and semi-arid countries agriculture is often confined to irrigated land, with very little farming possible in non-irrigated or rainfed areas. Agricultural land constitutes only a part of any country's territory, which in addition also includes areas not suitable for agriculture, such as forests, mountains, and inland water bodies. Agricultural land covers 38% of the world's land area, with arable land representing less than one-third of agricultural land (11% of the world's land area). In the context of zoning, agricultural land (or more properly agriculturally zoned land) refers to plots that may be used for agricultural activities, regardless of the physical type or quality of land.

SEED TREATMENT

In agriculture, a seed treatment or seed dressing is a chemical, typically antimicrobial or fungidal, with which seeds are treated (or "dressed") prior to planting. One such seed dressing, Imidacloprid, from the Neonicotinoid family of insecticides, is controversial and has been banned in France, due to that government's belief that the chemical is implicated in recent dramatic drops in bee counts, and possibly in the so-called Colony Collapse Disorder. In order to qualify for the United States Department of Agriculture Organic certification, farmers must seek out organic seed. If they cannot find organic seed, they are allowed to use conventional, untreated seed. Treated seed however, is never allowed.

Organic Seed Treatment Notes

Preventing Disease in the Field

An ounce of prevention is worth a pound of cure. Organic seed disease prevention starts with health-promoting cultural practices in ecologically managed farms that prevent disease and pests in the first place, that include:

1. Crop rotation- creating disease-suppressive soil/compost, habitats for benefical pest predators
2. Appropriate Planting Dates, Soil Temperature and Moisture
3. Selecting Disease-Resistant Varieties
4. Cleaning and Processing Methods that Control Disease

Seed Surface Treatments

Surface seed treatments reduce disease-causing fungi and bacteria found on the seed. Biological seed treatments control seed pests by parasitizing the pest organisms, competing for food on the root system, or producing toxic compounds that inhibit pathogen growth. Control of surface pathogens include beneficial microbes in compost teas, herbal sprays, washes or oils, hot water, heat, and disinfectants.

1. Compost and Vermi-Compost Teas
2. Biodynamic Treatments

3. Herbal Treatments
4. Hot Water Bath
5. Disinfectants
   (a) Commercial Products
6. Indigenous Methods
7. Research and Resources

Treatments to Prevent Soil-Borne Disease

This approach surrounds the germinating seed with protection from the disease-causing fungi, soil-borne pathogens that cause damping-off, and other soil-borne diseases. Fungal pathogens in the soil can rot the seed before it emerges from the soil or kill plants as they emerge. Treatments encourage healthy root systems.

1. Drench Transplant with Compost Tea - *inoculant to enhance beneficial soil microflora*

Research and Resources

Seed Surface Treatments

Compost Tea and Vermi-Wash

Disease-suppressive compost has complex microbial communities that compete with and control pathogens. It is typically used to coat the leaf surface with beneficial microbes or as a soil drench. To produce suppressive compost, we suggest to pre-mix the raw ingredients, manage at lower temperatures on fertile soil, minimize turning and inoculate with earthworms to increase beneficial microbes. Vermicompost (earthworm castings) has been found to have significant disease-suppression. Research is needed to investigate use of compost teas for control of seed-surface pathogens.

2

Formation of Soil

SOIL FORMING FACTORS

Earth Deposits: A Basis for Creating Landforms and Soil

Physical and chemical weathering processes erode parent material into mineral particles and dissolve minerals in solution, but weathering does not transport or deposit weathered minerals. Volcanoes, wind, water, ice, and waves transport and deposit weathered minerals, they are influenced by gravity, and upon deposition create landforms from which soil formation begins. Minerals weathered from parent material determine the mineral content in sand, silt, and clay.

Sand: The major mineral in sand is called quartz and it is composed of silica and oxygen (SiO_2). Sand grains vary in size depending on how long they are exposed to weathering.

Quartz is very resistant to weathering; therefore, sand grains are larger in diameter compared to silt and clay particles:

- *Very coarse sand:* 2.00 - 1.00 mm diameter.
- *Coarse sand:* 1.00 -.50 mm diameter.
- *Medium sand:* 0.50 - 0.25 mm diameter.
- *Fine sand:* 0.25 - 0.10 mm diameter.
- *Very fine sand:* 0.10 - 0.05 mm.

Because sand is larger in diameter than other soil separates, sand grains provide larger spaces in which water and air can more easily move through soil. Sandy soils are limited in nutrients because the nutrients leach out from the large pore spaces between sand grains, so crops are generally not grown in sandy soils.

Silt: Silt contains silicate minerals like sand but the diameter of silt particles is smaller, 0.05 - 0.002 mm, and so the pore space between silt particles are smaller. Therefore, silt has the ability to hold water between particles and retains nutrients for plant use. Silt is an ideal soil for growing crops.

Clay: Silicates, mica, iron, and aluminum hydrous-oxide minerals are found in clay. The silicate clay group is primarily located in the mid-latitudes, while the iron and aluminum clays are found in the tropic zones. Clay particles are 0.002 mm in diameter or smaller, so the pore spaces between clay particles are very small.

Thus, water and air movement through clays particles is significantly decreased. When clay becomes wet it swells, sticks together (cohesion), and feels 'sticky'. As wet clay dries it shrinks and cracks. Clay also becomes dense, hard, and brittle making it difficult for plant roots to grow through. Clay soils containing laterite and smectite properties are not desirable to grow crops in or to build on.

Alluvial Deposits

Rock debris that has been eroded into fine sediments that are subsequently transported by a mountain stream or river to the valley floor, as the gradient of the mountain decreases. Sediment is carried by either ephemeral (intermitant) water flow that occurs in arid climates or perennial stream water flow that occurs in humid climates, and is subsequently distributed into fan shaped landforms called alluvial fans.

Alluvial fans are known to be either wet or dry, depending on if the fans are located in humid or arid climates. Alluvial soils are finely layered and are very deep. Closer to river banks and on natural levees alluvial soils are more sandy, however alluvial soils are more clayey to peaty when closer to swampy areas. Alluvial deposits such as the Mississippi River Delta and China's vast alluvial plains have rich top soil and are known for being very fertile, crop growing regions.

Colluvial Deposits

Materials that move downslope by force of gravity and/or erosion and collect at the base of mountains or foothills, with little or no sorting. Talus cones are a type of colluvial deposit. Soils from colluvial deposition are generally deep and fragipans (hard clay soil) are common.

Eolian Deposits

Eolian deserts form in arid regions of the world where dry air masses create wind systems that transport and deposit loose sediments. Silt particles, called loess, are carried by wind even longer distances than sand and collect around the fringe of deserts. Large areas of the desert environment that receive more than 125 square kilometres of eolian sand are called sand seas or ergs, such as Erg Chech in Algeria. The largest desert in the world, the Sahara Desert, is 7 million square kilometres and contains several ergs. Smaller areas are called dune fields. Wind force and variable wind directions transport and deposit sand and in the process create different types of dunes. Some dunes are shaped by the wind into ridges, strings, domes, stars, or barchans (half-moon shaped dunes). Deserts primarily consist of wind-deposited sand which originated from sandstone that eroded over time.

Glacial Deposits

Glaciers are large and small ice masses that are found at high latitudes on Earth.Mountains located at all latitudes have small glaciers. During the Pleistocene, 10,000 years ago, glaciers extended into much lower latitudes and elevations than are currently located. As the climate changed and weather got warmer, glaciers began to melt and abrade bedrock lying below the glaciers. Varying rates of ice melt caused eroded sediment to 'drop out' of retreating, melting glaciers. This 'glacial till' formed deposits called moraines and drumlins. Glacial till consists of unstratified (unlayered) and unsorted glacial deposits, some the size of huge boulders.

Meltwaters flowing upon, under, within or at the margin of glaciers accumulate deposits known as outwash plains and kettles (depressions), kames (small, mound shaped accumulations of sand or gravel), and eskers (narrow, sinuous ridges of sediment).Where glaciers extend beyond the mouths of river valleys and enter the sea, their glaciomarine sediment load is dumped into the ocean. As climates warm glaciers melt and retreat. Glaciofluvial (glacier stream water) sediment is transported downstream by way of glacial meltwater and is deposited in braided streams. Glaciolacustrine (glacier lake water) sediment is deposited in glacial lakes when damming of ice or moraines occurs, and fluctuations of meltwater flow create distinctive varve deposits. Fine glacial debris consisting of silt and clay becomes airborne where vegetation is not present to hold this sediment down, and often travelling hundreds of kilometres before landing and forming loess deposits. The Muir Glacier and Margerie Glacier in Glacier Bay, Alaska are actively retreating glaciers.

Lacustrine Deposits

Lakes are different than marine environments in that sedimentation of lakes is ten times higher than in marine environments. Lakes are also smaller, are nearly closed systems, and tides in lakes are less pronounced. Therefore energy levels in lakes are lower, coarser sediment (sand and gravel) is deposited in shallow water areas of lakes, especially during summer, while finer-grained sediment (silt and clay) is deposited in deeper water areas of lakes, and more so during winter. Varves, alternating thin layers of light-coloured coarser grained sediment and dark-coloured finer grained sediment, are one type of lacustrine deposit and form in both glacial and nonglacial lakes.

Deposits in open lakes come mainly from rivers but may also be deposited by wind, ice-rafting, and volcanic rock erosion. Sedimentation in closed lake systems consists of evaporite minerals, carbonate muds, sands, and silts. Lacustrine deposits are often rich in organic shales which are important source rocks for petroleum. Well-known lacustrine shale deposits in the world include the Eocene Green River Formation in Wyoming, Utah and Colorado; the Jurassic Morrison Formation of the Colorado Plateau; Devonian sediments from the Old Red Sandstone of the Orcadian Basin in northeast Scotland; and the Triassic Keuper Marl of South Wales, just to name a few.

Loess Deposits

Loess is comprised primarily of silt grains, with less significant anounts of clay and sand. The mineral quartz is most dominant in loess with feldspars, carbonates, and clay minerals present in smaller amounts. For instance, in arid regions loess contains larger amounts of calcium carbonate; whereas, in humid regions clay minerals in loess are more prevalent.

Desert regions of the world may be thought of as prime locations for loess deposition because of the availability of loose sediment, sparse vegetal cover, and moderate to strong winds. However, loess deposits are more commonly located in or near glacial regions.

Glacial outwash debris containing sand, silt, and clay is transported to floodplains by rivers that drained glacial meltwater. The glacial debris, primarily the silt and clay, becomes airborne via strong winds as vegetation is not present to hold sediment down. Loess can sometimes become suspended several kilometres high and hundreds of kilometres in distance, with tens to hundreds of tons of sediment being transported in a single 'dust storm', as was the case in the 1935 dust storm over the midwest United States. Near Wichita, Kansas a dust storm had suspended about five million tons of sediment over a 78 square kilometer area and around 300 tons per square kilometer of dust was deposited from the same storm near Lincoln, Nebraska.

Marine Deposits

Physical processes mainly rework and distribute carbonate materials on marine shelf but can also help in the production of carbonates. Moderate water ciruculation on marine shelf brings nutrients from deeper water to shallow shelf region which aids in organic growth of ooids, fecal pellets that eventually become cemented together. Waves constantly move fine carbonate mud and coarser sediment to form sand or gravel covered tidal flats, beaches, dunes, marshes, lagoons, and swamps or transports these sediments seaward to form spits, tidal deltas and bars, and barrier islands.

Waves pounding against coastal rocks also contribute rock particles and sediment to the coastal shelf and seaward. The Outer Banks of North Carolina are a chain of barrier islands that contain beaches, lagoons, and spits. Reefs can be characterized as either thick masses of living carbonate 'rock', or structures produced by sediment-binding, live organisms. The Great Barrier Reef, Australia, is the largest coral reef in the world.

Bioherms are mounds of dead organic material that have collected in rocks of different composition. Organisms are able to extract calcium carbonate ($CaCO_3$) from seawater to build protective shells or skeletons, although the availability of $CaCO_3$ in seawater is controlled by pH, temperature, and carbon dioxide content. When these organisms die their remains collect and form carbonate deposits known as bioherms. Carbonate formers in Earth's current oceans are not the same as those that formed carbonates in ancient oceans.

Other marine depositional environments include deltas, beaches, barrier bars, estuaries, lagoons, and tidal flats. Beach and barrier bar deposits are mostly contain fine to medium grained, well sorted sand as well as placer gold, platinum, and other minerals.

Estuarine deposits, like in the Chesapeake Bay and San Francisco Bay, consist of cross-bedded sands and mud, or a mixture of both sand and mud. Lagoonal deposits include evaporites, fine-grained sediments, and black shales. Delta deposits and tidal flat deposits, like the Mississippi River Delta, primarily contain muds in the upper zone, mud and sand in the middle zone, and sand in the lower zone.

Volcanic Deposits

Volcanoes produce magmas consisting of various mineral compositions that in turn create various rock types. The amount of gas in magma and the viscosity (thickness) of magma determine the volatility of a volcanic eruption and the types of landforms that are formed. Continents and oceanic environments contain highly fluid, basaltic magma whereas magma forming as island arcs at the margins of some continents consists of high silica lavas that are more viscous and crystallize into rhyolites, andesites, and dacites. Lava in orogenic (mountain building) environments is most viscous (thick) and has a higher gas content so eruptions are more explosive and form an extrusive, solid volcanic material called tephra. Volcanic ash is found in the United States primarily in Hawaii, Washington, Oregon, and also in Japan, Indonesia, Central America, and other mountainous regions of the world. Most volcanic ash forms into very fertile soil that is used for growing crops.

Three major volcanic landforms are created as a result of volcanic activity.

1. *Lava Plains and Plateaus* : These types of volcanic landforms are created as a large volume of fluid lava flows over an expansive surface area. The resulting topography are extremely flat surfaces that aggrade with each successive lava flow that is mafic in composition. An example of a lava plain is located in the Columbia River Plain in Washington and Oregon. Oceanic plains and plateaus can also form, even more extensively, from this type of lava flow.
2. *Cones or Shields* : Most volcanoes form into *composite cones* that have a distinctive appearance of layers of interbedded, blocky lava with tephra that is mostly ash or cinder. The peaks of composite cones have narrow, circular bases whose peaks rise several thousands of meters. Mount Ranier, in Washington is a composite volcano. Conversely, shield volcanoes are comprised of fluid basaltic magma with very little tephra and therefore have lower peaks than composite cone volcanoes. The chain of Hawaiian volcanoes are shield volcanoes.
3. *Calderas:* This type of volcanic landform is created once eruptions occur and subsequently the upper part of a volcano collapses inward. Volcanoes containing tephra sheets, such as the composite cones, are more prone to forming calderas once an eruption occurs. Crater Lake, Oregon and Yellowstone Plateau,Wyoming are calderas.

Landforms are not always formed by deposition alone. For instance, the spectacular Grand Canyon land formation has been cut and etched by the erosive forces of wind and water over millions of years. The Colorado River has deeply cut gorges into less resistant rock and created a canyon while more resistant, less weathered rocks, such as sandstones, give the Grand Canyon its statuesque appearance. Also, human activity on the land can have a significant affect on erosion due to construction and agricultural practices, as is seen in this picture of eroded sediment entering and filling the San Francisco Bay, in California.

FERTILITY OF SOIL FORMATION

Soil fertility is greatly influenced by the factors of soil formation. As soils form, nutrients are being continually removed from and added to the soil with time. The conditions that are present during soil formation ultimately determine how much and what kind of nutrients the soil can naturally supply and hold. In the previous section, we discussed how soil profiles vary from place to place. In fact, differences in soil make-up cause soils to behave and perform in many different ways. We have mentioned the process of weathering when we discussed soil mineral composition. Now, we will discuss the processes of weathering in terms of soil formation. We will begin by looking at the 5 factors of soil formation that cause great differences in the development of soil profiles.

Parent Material

Parent material affects soil fertility in many ways. First, the type of parent material determines which minerals (link to mineralogy) will predominate in the soil. Secondly, as parent material weathers, nutrients are released into soil solution, which subsequently can be taken up by plants and other organisms or leached from the soil.

Parent material is the starting point for most soil development. The parent material may be mineral rock and/or organic matter. When parent rock material is exposed to the atmosphere or when organic matter and/or minerals are deposited on the earth's surface, soil formation begins. The type of parent material and how the soil is formed will greatly influence the properties of the soil.

There are two general rules that we may state about the influence of parent material on soil formation:

- Finely textured parent materials tend to weather into finely textured soils. Coarsely textured parent materials tend to weather into coarsely textured soils.
- Darker minerals tend to yield more fertile soils.

Soil Texture and Soil Structure

Soil texture and soil structure are both unique properties of the soil that will have a profound effect on the behavior of soils, such as water holding capacity, nutrient retention and supply, drainage, and nutrient leaching.

In soil fertility, coarser soils generally have a lesser ability to hold and retain nutrients than finer soils. However, this ability is reduced as finely-textured soils undergo intense leaching in moist environments.

Soil Texture

Soil texture has an important role in nutrient management because it influences nutrient retention. For instance, finer textured soils tend to have greater ability to store soil nutrients. In our discussion on soil mineral composition, we mentioned that the mineral particles of a soil are present in a wide range of size. Recall that the fine earth fraction includes all soil particles that are less than 2 mm. Soil particles within this fraction are further divided into the 3 separate size classes, which includes sand, silt, and clay. The size of sand particles range between 2.0 and 0.05 mm; silt, 0.05 mm and 0.002 mm; and clay, less than 0.002 mm. Notice that clay particles may be over one thousand times smaller than sand particles. This difference in size is largely due to the type of parent material and the degree of weathering. Sand particles are generally primary minerals that have not undergone much weathering. On the other hand, clay particles are secondary minerals that are the products of the weathering of primary minerals. As weathering continues, the soil particles break down and become smaller and smaller.

Textural Triangle

Soil texture is the relative proportions of sand, silt, or clay in a soil. The soil textural class is a grouping of soils based upon these relative proportions. Soils with the finest texture are called clay soils, while soils with the coarsest texture are called sands. However, a soil that has a relatively even mixture of sand, silt, and clay and exhibits the properties from each separate is called a loam. There are different types of loams, based upon which soil separate is most abundantly present. If the percentages of clay, silt, and sand in a soil are known (primarily through laboratory analysis), you may use the textural triangle to determine the texture class of your soil.

The major textural classes for the soils. Each of the textural classes make up finely textured soils. As you can see, soil surveys show that more than 90% soils are finely textured. This is largely due to the type of parent material of most soils, which is basalt. Since basalt is a finely textured rock, it weathers into finely textured soils. The relative amount of clay has great importance in the soil.

Importance of Clay and Other Particles of Similar Size

Clay particles, as well as other particles of similar size, are important components of a soil. There is a fundamental difference between soils that contain large amounts of sand particles and soils that contain large amounts of very small particles, such as clay. That difference is surface area. The total surface area of a given mass of clay is more than a thousand times the total surface area of sand particles with the same mass. To put this idea into perspective, imagine a single cube with 6 sides. This cube represents a sand particle. Now, imagine that you break this single cube up into 100 smaller cubes, which represent 100 clay particles. These 100 cubes each have 6 sides. Essentially, by breaking up the larger cube, you have exposed many more surfaces. Thus, the total surface area of the smaller cubes will be much greater than the surface area of the single cube. This increase in surface area has an important implication in nutrient management because it provides many places for soil particles to retain and supply nutrients (such as calcium, potassium, magnesium, phosphate) and water for plant uptake

Types of Very Small Particles within the Soil

- The most common clay minerals in soil are called layered silicate clays, or phyllosilicates. There are different types of layered silicates, such as kaolinite, halloysite, montmorillonite, and vermiculite. The various types of layered silicates differ greatly, as we will discuss later.
- Amorphous minerals, such as allophane, imogolite, and ferrihydride, may be found in the volcanic soils of that developed from volcanic ash. Like silicate clays, these minerals have a very high surface area. As a result, soils with amorphous minerals hold large amounts of water and stored nutrients, depending on the degree of weathering.
- Aluminum and iron oxides are typically found in the highly-weathered soils of the tropics. As clay minerals are intensely weathered, the structure of silicates clays change. Particularly, the silicate clays lose silica. What remains in the soil are aluminum and iron oxides. Gibbsite is an example of an aluminum oxide, which has a grayish, whitish hue. Goethite is an example of an iron oxide, which imparts a reddish color to the soil.

Properties of Oxides

- Oxides are fairly stable and resistant to further weathering.
- Oxides can act like a glue and hold other soil particles together.
- Oxides can tie up nutrients, such as phosphorus.
- Oxides have a high anion exchange capacity (AEC).
  - Humus is the portion of organic matter that is mostly resistant to decomposition and remains in the soil. Humus is composed of small particles, with tremendous surface area. These particles have a very great capacity to retain and supply nutrients, as well as hold water.

Soil Structure

Soil structure is the arrangement of soil particles into groupings. These groupings are called peds or aggregates, which often form distinctive shapes typically found within certain soil horizons. For example, granular soil particles are characteristic of the surface horizon. Soil aggregation is an important indicator of the workability of the soil. Soils that are well aggregated are said to have "good soil tilth."

Soil Aggregates

Generally, only the very small particles form aggregates, which includes silicate clays, volcanic ash minerals, organic matter, and oxides. There are various mechanisms of soil aggregation.

Mechanisms of Soil Aggregation

- Soil microorganisms excrete substances that act as cementing agents and bind soil particles together.
- Fungi have filaments, called hyphae, which extend into the soil and tie soil particles together.
- Roots also excrete sugars into the soil that help bind minerals.
- Oxides also act as glue and join particles together. This aggregation process is very common to many highly weathered tropical soils and is especially prevalent.
- Finally, soil particles may naturally be attracted one another through electrostatic forces, much like the attraction between hair and a balloon.

Aggregate Stability

Stable soil aggregation is a very valuable property of productive soils. Yet, the stability of soil aggregation is very reliant on the type of minerals present in the soil. Certain clay minerals form very stable aggregates, while other clay minerals form weak aggregates that fall apart very easily.

- Highly weathered silicate clays, oxides, and amorphous volcanic materials tend to form the most stable aggregates. The presence of organic matter with these materials improves stable aggregate formation. In nutrient management, the aggregate stability is important because well-aggregated minerals are well drained and quite workable.
- In contrast, less weathered silicate clays, such as montmorillonite, form weak aggregates. Some silicate clays are said to have a shrink-swell potential. This means that the soil minerals expand, or swell, when wet, causing the soil to become sticky and drain poorly. When dry, these soils shrink and form cracks. The make-up of the lattice structure of silicate clays determines the shrink-swell potential. Although there are no soils with a shrink-swell potential, these soils may be found on Molokai.

Soil Mineralogy

Soil mineralogy is closely related to soil fertility. Differences in soil mineralogy cause great differences in soil fertility. For instance, moderately weathered clays attract and retain greater amounts of nutrients than highly weathered clays and oxides. Though volcanic ash inherently has a low ability to hold nutrients, it interacts with organic matter to produce very fertile soils. Knowledge of mineralogy helps to determine the appropriate nutrient management strategy for your soil.

In our discussion on soil texture and structure, we mentioned that the very small particles form aggregates. The major groups of small particles include silicate clays, organic matter, volcanic ash minerals, and oxides. In fact, this grouping of small particles is also used to describe the distinct categories of soil mineralogy (types of clay minerals).

Soil mineralogy:

- Layered silicate clays
  - High activity
  - Low activity
- Organic matter
- Volcanic ash material
- Oxides

Soil behavior is greatly influenced by these types of clay minerals and/ or the amount of organic matter that a particular soil type contains.

Before discussing differences in soil mineralogy, it is very helpful to understand the concepts of cation and anion exchange capacities (CEC and AEC). CEC and AEC are properties that can help differentiate soil minerals.

Cation Exchange Capacity (CEC) and Anion Exchange Capacity (AEC)

CEC and Less Weathered Soils

Less weathered soils, that contain minerals such as montmorillonite, are said to have a 'cation exchange capacity,' or CEC, under acidic, neutral, and alkaline conditions. CEC is the soil's ability to attract, retain, and supply nutrients, such as calcium, potassium, ammonium, and magnesium. These nutrients are positively charged atoms, known as cations. The surfaces of less-weathered clay minerals (such as montmorrillonite) generally have a negative charged. Much like a magnet, negatively charged soil surfaces attract positively charged cations. However, under acidic conditions, the soil will also have a tendency to attract aluminum and hydrogen cations. The presence of aluminum and hydrogen contributes to soil acidity. In contrast, under alkaline conditions, the soil attracts sodium which contributes to soil alkalinity.

AEC and Highly Weathered Soils

Although the majority of the world's soils have CEC, the highly weathered soils of the tropics are an exception. In addition to the having CEC, many tropical soils also have an 'anion exchange capacity,' or AEC, depending upon the pH of the soil. Under neutral and alkaline conditions, the soil has CEC, like the less weathered soils. However, under acidic conditions, these soils generate AEC. This means that the soil becomes positively charged and attracts, retains, and supplies negatively charged anions, such as sulfate, phosphate, nitrate, and chloride. For soils with AEC, proper management of pH is crucial in order to provide sufficient amounts of the nutrient cations (calcium, magnesium, ammonium, and potassium).

Minerals that exhibit AEC are highly weathered kaolinite, aluminum and iron oxides, organic matter, and the allophanes and imogolites of volcanic soils. Highly weathered Ultisols and Oxisols, volcanic Andisols, and organic Histosols all have AEC under acidic conditions. The pH at which these soils develop AEC differs depending upon the minerals within the soil. Since organic matter only generates AEC at a very low pH, it is still a good source of CEC.

Importance of CEC and AEC

CEC and AEC values are important measurements that provide us with important information regarding the soil's ability to retain and supply certain nutrients to the plant. In addition to nutrient retention, CEC and AEC helps us predict the leaching potential of certain nutrients in areas with high rainfall. When the soil has a very high CEC, negatively charged nutrients such as nitrate are not be retained by the soil. Instead, nitrate leaches through the soil profile in areas with high amounts of precipitation. Likewise, soils with high AEC experience leaching of positively charged nutrients, such as calcium and potassium.

CEC is often expressed in centimoles of charge per kg of soil. By sending a sample of your soil to a soil testing laboratory, you can determine your soil's CEC. The value of knowing a soil's CEC cannot be underestimated in nutrient management. Without it, your soil would not be able to provide your plants with sufficient amounts of nutrients. However, one must be cautious when interpreting the laboratory results for CEC depending on the soil type being analyzed. You can obtain accurate results for most laboratory methods that measure CEC in less weathered soils with permanently negative charge. On the other hand, these laboratory methods can overestimate the CEC of highly weathered soils that have an AEC. This is because the pH of highly weathered soils affects the CEC of the soil. As the pH of highly weathered soils with AEC increases, the CEC of the soil also increases. Thus when the method used to determine the CEC uses solutions that raise the pH of the soil, the reported CEC is higher than its actual value in the field.

Layered Silicate Clays

Layered silicate clays are secondary minerals that have formed as the result of weathering of parent material. There are two major categories of layered silicate clays within the soil: high activity clays and low activity clays.

High Activity Clays

Generally, soils with large amounts of high activity clays are not highly weathered. High activity clays have a high 'cation exchange capacity' (CEC), due to their large surface area. This means that these clays have a great capacity to retain and supply large quantities of nutrients, such as calcium, magnesium, potassium, and ammonium. Not only do these clays have a large CEC, but they will generate CEC under all soil conditions regardless of soil pH. As a result, these clays tend to produce highly fertile soils. Examples of these clays are montmorillonite (and other smectites), vermiculite, illite, and mica.

Cation Exchange Capacity or Anion Exchange Capacity

- High activity clays have a cation exchange capacity (CEC). Although high activity clays will not have anion exchange capacity (AEC), the CEC increases as pH increases and decreases as pH decreases.

Montmorillonite

Some high activity clays, such as montmorrillonite, have a shrink and swell potential. This means that the clays will shrink and crack when dry, and expand and swell when wet. With little additions of nutrients, these soils may be very productive. However, the shrink and swell potential will result in poorer drainage. And so, proper management of irrigation is required.

Low activity clays

In contrast, low activity clays are more highly weathered. Thus, due their lesser surface area, low activity clays have a lower capacity to retain and supply nutrients. In addition to CEC, low activities clays can also have AEC, depending upon the pH of the soil. The AEC causes these clays to retain and supply nutrients, such as phosphate, sulfate, and nitrate, rather than the base cations, under acidic conditions. Yet, under neutral and alkaline conditions, these low activity clays generate a CEC.

Cation Exchange Capacity or Anion Exchange Capacity?

- Under acidic conditions, low activity clays have an AEC
- Under neutral and alkaline conditions, low activity clays have a CEC

Management of pH

In order to provide adequate amounts of the base cations, proper management of pH is crucial. If soil pH is low, additions of lime and/or organic matter may increase the CEC for soils high in low activity clays.

Low activity clays have a low shrink and swell potential. With additions of nutrients, these soils may be very productive soils.

Organic Matter

Most soil organic matter accumulates within the surface layer of the soil. This organic matter may be divided into two groups: non-humic matter and humic matter.

Non-humic matter includes all undecomposed organic material within the soil. Examples of non-humic matter are twigs, roots, and living organisms.

Humic matter includes humic acids, fulvic acids, and humin. (Humin is the dark material in soil that is highly resistant to decomposition.)

Importance of soil organic matter

- Due to its tremendous surface area, soil organic matter:

–Acts like a sponge to store water

–Retains and provides nutrients (CEC)

–Glues and binds soil particles into stable aggregates

- Reduces the occurrence of aluminum toxicities.

Like low activity clays, organic matter may have either CEC or AEC, depending upon soil pH. However, it will rarely have AEC. In fact, the pH must fall to approximately 2.0 before it will have AEC.

Cation Exchange Capacity or Anion Exchange Capacity?:

- Soil organic matter may have both AEC and CEC. However, the charges on organic matter are dependent upon soil pH. For soil organic matter to generate an AEC, the soil pH must be 2.0.

Management

Without additions of organic matter, tillage practices will greatly reduce organic matter content in the soil. And so, no-till and minimum tillage systems with the return of organic matter to the soil are gaining favor by farmers to improve and conserve soil quality.

PROCESSES OF SOIL FORMATION

Soil production occurs by a variety of processes through the addition, loss, transformation, and translocation (movement) of material within a soil profile. These processes include the accumulation of organic matter at the ground surface and its subsequent integration down into the soil, decomposition of primary minerals into secondary minerals, the leaching of particular organic and mineral constituents from the parent material, and the movement of material leached or washed from upper soil layers deeper into the soil.

Additions, Losses, and Transformations

New material may be added to a soil profile from sources below, within, or above the ground surface. The breakdown of bedrock adds mineral matter to soil from below. Organic matter is added to soil profiles from above when leaves and other organic debris fall to the ground and the remains of once-living plants and animals become mixed into the soil, or roots die within the soil. Water infiltrating into the ground deposits dissolved matter within the soil profile in response to significant changes in pH or solute concentration (such as through evaporation). In addition, floods can deposit fluvial sediment on floodplains and wind can add dust to the top of a soil profile. Soil profiles lose material by either erosion or dissolution. Erosion by surface water flow and wind removes material from the top of the soil profile, and dissolution removes mineral material from within the profile. Soil-forming material is transformed by chemical weathering processes that convert primary rock-forming minerals into secondary minerals, and the breakdown of organic matter into humus.

Soil scientists refer to the movement of material from one part of a soil profile to another as translocation. Dissolved matter may move within the soil or leave the soil in groundwater, a process known as leaching. Water percolating down through the upper soil may also physically entrain clays, silt, and colloids and re-deposit them deeper in the soil. Precipitation infiltrating into a soil dissolves material from the upper layers. Soil water flow carries it downward and may concentrate it in lower layers. The process of transporting clays and mobile ions out of a soil horizon is called eluviation; washing of material into a soil horizon is called illuviation. The formation of caliche, a hard carbonate-cemented layer, is an important process in arid region soils where infiltrating water evaporates within the soil and deposits the dissolved material it was carrying. Over time such deposition can build thick layers of erosion resistant carbonate hardpan. Cumulic soils are those that accumulate slowly enough, or episodically enough, that material added at the top of the soil profile weathers to some degree before being buried. Examples of such inputs include wind-blown dust or volcanic ash and flood-deposited silts on floodplains. Fine-grained cumulic soils are particularly vulnerable to wind erosion if stripped of their vegetation cover, such as happened to the fertile silt (loess) originally deposited by wind across the Great Plains of North America in glacial times that blew away when plowed during the Dust Bowl era drought in the 1930s. Addition of wind deposited dust may also significantly alter soil properties. For example, incorporation of dust infiltrating down into a permeable soil may eventually clog pore spaces and decrease permeability enough to alter the intensity and pace of chemical weathering.

FACTORS OF SOIL FORMATION

Characteristics of a soil are determined by five factors of soil formation:

- The composition of the parent material—the material from which the soil formed;
- The climate under which the soil formed;
- The organisms or plant and animal life on and in the soil;
- The relief or shape of the land surface; and
- The length of time the forces of soil formation have been acting on the soil material.

Parent material and topography are the factors of soil formation present when the soil starts to develop. Climate and plant and animal life then act on these initial conditions to cause formation of soil horizons. Finally, time is needed for changing parent material into soil.

A soil profile, as seen in a road cut or pit, is a vertical exposure of the various horizons or layers, including the topsoil, subsoil and substratum or undeveloped parent material. Distinct soil horizons develop as a result of the influence of these five soil formation factors. This usually takes hundreds or thousands of years.

Parent Material

The parent material is the starting point in soil formation. It largely determines the range of texture (*i.e.*, per cent composition of sand, silt, and clay) that a soil will have. Its initial chemical and mineralogical properties will affect the natural fertility of the soil formed.

Indiana soils have many kinds of parent material. They include material that weathered from bedrock and mainly stayed in place and material that was carried some distance by ice, water, or wind. Glaciers (large ice sheets) transported the parent materials of many Indiana soils. The map of the soil regions of Indiana, is based mainly on soil parent material. Guidelines for identifying rocks and parent materials, Soil and Landscape Properties.

Before the glacial period, the topography of the state was rather rough and rolling, similar to that now found in southern Indiana (Soil Regions 11, 12, and 13), which was never covered by the ice. Glaciers, ice sheets hundreds of miles long and hundreds to a few thousand feet thick, covered most of the state during at least three different ice ages. From the oldest to the youngest, these glacial ice ages have been called the Kansan, the Illinoian, and the Wisconsinan Ages. As the ice moved southward, it destroyed old hills and made new ones. The material carried by the ice buried old valleys. A layer of rock, sand, silt, and clay was left when the ice melted and receded. This material—collectively called glacial drift—is partly till and partly outwash. Till is a homogeneous mixture of sand, silt, and clay, and usually gravel and boulders, that was deposited directly from the ice of the glacier. It makes up much of the parent material in Soil Regions 6, 7, 8, 9, and 10. In the Miami, Crosby, and Parr soils (Plates 2, 3, and 5, in the centerfold section), the C horizon is glacial till. Outwash was deposited by running water from melting glaciers. The size of the particles that make up outwash varies according to the velocity of the water that carried them. When fast moving water slowed down, coarser particles were deposited as outwash, but the finer particles, such as fine sand, silt, and clay, continued to be carried by the slower moving water. Outwash deposits generally consist of layers of sandy and gravelly material. In soils formed from outwash, typically the substratum (C horizon) consists of coarse sand and gravel, and the subsoil (B horizon) is finer in texture, as in the Ockley soil (Plate 6). Outwash makes up much of the parent material of Soil Region 3. Lacustrine (lake bed) materials were deposited from slowly moving or ponded glacial meltwater. Because the coarser particles dropped out of moving water as outwash, only the finer particles, such as fine sand, silt, and clay, remained to settle out of still water. Soil Regions 1 and 2 contain many lacustrine areas. Those in northern Indiana were deposited during the most recent glacial period (Wisconsinan).

Many lacustrine deposits in southern Indiana are older. As the ice receded, there were periods when the flood plains of the glacial meltwater rivers became dry. During these periods, wind picked up silt and sand particles from the flood plains and carried them towards the upland. Fine and medium sand particles were carried only a short distance, then deposited in sand dunes close to the flood plain. Eolian sand (wind-blown) is an important soil parent material along the major rivers of southern Indiana, near Lake Michigan and in the Kankakee Valley—Soil Regions 1 and 4. The parent material of the Chelsea soil is eolian sand (Plate 7). Silt was carried by the wind a greater distance from the flood plain and then settled out of the air as loess. Loess was blown mainly from the Wabash and White River Valleys in central and southern Indiana. Near the river valleys, in Soil Region 5, loess is thick, and the soils formed in it are deep. The Hosmer soil (Plate 12), found in Region 5, formed in more than five feet of loess. Farther away from these valleys, loess is thinner but still is the parent material of the upper part of the soil. For example, the upper horizons of most of the soils in Regions 8, 9, and 10, and of some of the soils in Regions 11, 12, and 13 are in loess. South central Indiana was never covered by glacial ice. The parent material in this area is weathered bedrock. The underlying sedimentary bedrock consists mostly of sandstone and shale in Region 11, limestone in Region 12, and limestone and shale in Region 13. Typical soils of these regions are the Caneyville soil (Plate 9) over limestone bedrock and the Trappist soil (Plate 10) over black shale. In several areas, weathered bedrock was covered with a layer of loess from a few inches to several feet in thickness. It is thicker near major rivers and becomes thinner away from the rivers. Loess is also thicker on nearly level uplands and thinner on sloping areas because it was eroded during and after the time of loess deposition. The variation in kinds of soils formed in these areas depends on the type of bedrock, the amount of loess that covers it, and the influence of other factors, such as relief and vegetation. The upper 18 inches of the Trappist soil (Plate 10) formed in loess.

Alluvium, sediment recently deposited by water, is the parent material of soil on flood plains. These materials are usually stratified (arranged in layers) The Genesee soil (Plate 8) is markedly stratified both in particle size and colour. The black colour is due to high organic matter content. Many soils on flood plains do not show this much stratification. Large areas of alluvial soils are on the flood plains along the major rivers of Soil Region 3. Smaller areas of these alluvial soils are also along streams in many other soil regions. A similar parent material is local overwash. It was washed into closed depressions (bowl-shaped landforms) from nearby higher areas, not carried in by a stream. Some soils formed in organic materials. After the glaciers melted, water was left standing in depressions in outwash, lake, and till plains. Grasses and sedges growing around the edges of these lakes died. These plant residues were preserved in the water and accumulated around the edges of the lakes. Later, white cedar and other water-tolerant trees grew on the areas. As these trees died, their residues also became a part of the organic accumulation. The lakes were eventually filled with a relatively non-decomposed organic material calledpeat. In peat, specific plant remains can be identified. Most peat has decomposed to muck, a black humus material in which the source plants cannot be identified. Organic soils in Indiana are mainly in Soil Region 1, with smaller areas in Regions 3, 6, and 7. Organic soils are not used in soil evaluation contests.

Relief of land surface

Relief describes the shape of the land surface. It influences natural drainage, movement of water in the landscape, erosion, plant cover, and soil temperature. Relief is described by landform and slope gradient (per cent) in soil evaluation. Natural soil drainage, erosion, and slope are also evaluated. Landform and natural soil drainage are related because the slope and shape of the soil surface determine which soils will lose water and which will gain water. Soils on steep slopes lose water and those in depressions gain water. As a result, soils on steep slopes are usually well drained regardless of the parent material, and those in depressions are usually poorly drained. Flat or gently sloping soils may be well drained if they are permeable and the water table is deep. However, if the water table is held up by slowly permeable material, the soils are wetter. The effect of natural drainage on soil development is evident in a comparison of soils that formed from similar parent material but under different drainage conditions. For example, the poorly drained Brookston soil (Plate 4) and the well drained Miami soil (Plate 2) both formed in loam-textured glacial till. Brookston soils are in depressions and have thick, dark surface layers and grayish subsoils with yellow brown spots or mottles. In contrast, Miami soils are on sloping areas and have yellowish brown colours in the upper part of the subsoil.

Regions 1, 2, 3, 7, 8, and 9 are, for the most part, nearly level but also include some sloping areas. Region 10 has nearly level areas broken up by steep slopes or ravines along drainageways. Soil Region 11 and Region 13 have a large proportion of steep slopes. Regions 4, 5, and 6 are more moderately sloping. In Region 12, soils overlie limestone bedrock. They are gently sloping and usually have many closed depressions, called sinkholes, that have no surface outlets. Water moves through the bottom of the sinkholes into underground caverns. Streams flow underground through these caverns. Water flows in surface streams only during storms.

3

Soil Organisms

Scientists have devised various schemes for characterizing and classifying soil organisms in order to be able to cope with their great diversity. A part from the conventional taxonomic approach, there are classification schemes based on body size, function (decompo-sition, etc.), and the role of organisms from an anthropocentric point of view.

The last-mentioned of these systems recognises 'productive' biota (such as crop plants), 'destructive' biota (pests, pathogens, and weeds) and 'resource' biota (species that contribute to soil processes such as decomposition but that do not produce a harvestable product). Under body size, there are three main groups: the microbiota (<100 μm diameter), the mesobiota (100 μm to 2 mm diameter), and the macrobiota (>2 mm diameter).

Microorganisms are the most abundant members of the soil biota. They include species responsible for nutrient mineralisation and cycling, antagonists (biological control agents against plant pests and diseases), species that produce substances capable of modifying plant growth, and species that form mutually beneficial (symbiotic) relationships with plant roots. This last group includes mycorrhizal fungi, various actinomycetes, and some bacteria. Within the soil biota, the most important groups of both destructive and resource organisms are the bacteria, fungi, nematodes, arthropods (such as mites and insects), earthworms (mostly beneficial), and weeds.

SOIL ORGANISMS AFFECT PLANTS

Soil organisms interact with plants in diverse ways. Some organisms increase plant growth, while others decrease it. Some alter root structure but others have little effect. Microorganisms can also have important influences on the way that roots branch and grow in soil. These structural modifications affect the uptake of water and nutrients by plants. Plant roots are of course the main structures through which soil organisms and plants interact. Roots provide a concentrated supply of carbon and energy for many heterotrophic soil microorganisms. In turn, soil animals that feed on fungi and bacteria multiply in response to the higher food supply around roots. Some organisms can invade plant roots.

Living roots are particularly important for many fungi, especially fungal pathogens and mycorrhizal fungi. Most root pathogens form associations with specific plants. Other pathogens colonise roots of a wide range of plants. Arbuscular mycorrhizal fungi generally form non-specific associations with many plant species. Ectomycorrhizal fungi are more specific in the types of roots they colonise. Before this section considers the effects of soil biota on plant roots, it first examines how plant roots shape the environment of soil organisms.

Plant Roots and Soil Biota

The rhizosphere is the region of soil that surrounds, and is influenced by, plant roots. A major influence in the rhizosphere is the release of organic molecules into the soil from roots. Although the exudates and mucilaginous materials released by roots diffuse through the soil, they always occur in their highest concentrations adjacent to the root surface. Because soil organisms are attracted to these exudates as a source of food, the abundance of soil organisms also increases close to roots. This is a dynamic relationship. The rhizospheres of old and young roots provide very different habitats for soil organisms.

As roots age, they release different types and quantities of carbon substrates, which in turn affect soil microorganisms. The rhizosphere also differs from the remainder of the soil in other ways. For example, the amount of inorganic nutrients can be either higher or lower in the vicinity of roots compared with that in the soil some distance away from roots. The concentration or depletion of nutrients around roots depends on physiological processes associated with root function, including nutrient uptake by plants. In addition, roots release hydrogen or bicarbonate ions into the soil, causing the pH in the rhizosphere to decrease or increase. Roots also consume or release oxygen. These contrasting effects, in addition to the considerable exudation of carbon compounds from roots into the soil and the compression of soil as roots grow in length and breadth creates a dynamic environment for soil organisms.

Some nutrients move towards roots in soil water by the process of mass flow. Other nutrients are less mobile because they are adsorbed onto soil particles. The availability of adsorbed nutrients such as phosphorus and copper to plants depends on the extent to which roots explore the soil and intercept these ions. This in turn affects the extent of the rhizosphere and the spatial distribution and activity of soil microorganisms. Water use by plants adds to the dynamic nature of the rhizosphere. Daily cycles of water uptake and loss by roots cause roots to expand and shrink. This adds to physical changes induced by seasonal cycles of wetting and drying or freezing and thawing. These daily and seasonal cycles alter the soil habitat, which influences the growth and activity of soil organisms. As a root develops, the number of organisms near the root surface increases rapidly. For many plants there is not a clear distinction between the root surface and the soil. Roots are dynamic, growing and casting off cells and navigating soil cavities and rough surfaces. As a consequence, the root surface can be highly irregular.

What is the Origin of Root Exudates?

A variety of organic compounds released from roots provide a source of energy and carbon for soil organisms. In a study of exudation from wheat roots, the plant was labelled with 14C. This technique allows the carbon released from the plant to be followed precisely. The greater the amount of 14C released into the rhizosphere of the wheat plant, the greater the quantity found in the microbial biomass. The quantity of the compounds released from roots is not the same along the length of a root. Older roots can be more 'leaky' than younger roots, although in this experiment there was no difference, at least for one line of wheat. The release of relatively large quantities of organic carbon into soil under normal conditions stimulates microbial activity and provides a source of energy that can accelerate the decomposition of organic matter.

How do Roots Affect Soil Organisms?

Roots provide an important habitat for bacteria, fungi and very small soil animals. The number of organisms can be about 500 times more in the rhizosphere than in the bulk soil. The surface of a root is not completely covered with bacteria and fungi, except in special circumstances, such as where roots are colonised by an ectomycorrhizal fungus. Usually less than 10 per cent of the root surface is colonised. Bacteria accumulate in crevices along the edges of outer root cells or where the root surface is damaged.

Some organisms form loose associations with roots and simply depend on the plant releasing molecules that they can use. Other organisms have more specific interactions with roots. For example, root nodule bacteria can stimulate the root to form additional cell tissue that proliferates to form a nodule. Root nodule bacteria live and multiply inside these root cells. When roots begin to disintegrate with age or through disease, there are more chances for opportunistic bacteria and fungi to enter through crevices between the surface cell layers or dead root cells.

Roots indirectly affect soil animals because they respond to changes in the abundance of bacteria and fungi associated with roots. The types of organisms present in the rhizosphere and in the bulk soil change over time (Lussenhop and Fogel 1991). These changes are associated with the intensity of grazing by soil animals and the abundance of other bacteria and fungi. Furthermore, there is a wide range in numbers of different types of animals. Overall, there is great diversity in type and number of organisms in the vicinity of roots.

How are Roots Affected by Soil Organisms?

Microorganisms can indirectly increase or decrease root growth. For example, organisms influence the movement of carbon from shoots to roots. This can alter root growth as well as stimulate the release of carbon into the rhizosphere. Some organisms increase root growth by producing compounds that function as root growth hormones. Organisms may decrease root growth by creating a significant drain on carbon reserves in the plant. This reduces the quantity of carbon in the root that would otherwise be used to form more roots.

The effects of soil organisms on roots can be either indirect or direct. For example, earthworms indirectly promote root growth because their burrows provide channels for roots to grow more deeply into the soil profile. In contrast, some rhizosphere bacteria and fungi impede root growth. Roots may grow more extensively if such microorganisms are absent than if they are present. Ectomycorrhizal fungi also reduce root growth by colonising the surface of roots and stunting their growth. In addition to the thickening of the root, the roots often branch in response to hormones produced by the fungus.

Arbuscular mycorrhizal fungi can reduce root growth. In a study of maize, microorganisms decreased both root hair length and root hair number. However, the extent of the decrease depended on whether arbuscular mycorrhizal fungi were present or not. In the presence of these fungi, both root growth and root hair number were considerably impeded in comparison to when these fungi were absent.

Generalisations about the effects of soil organisms on root growth are difficult to make because different species of plant and different types and combinations of soil organisms interact in different ways. Bacteria in the genus *Azospirillum* are interesting examples of microorganisms that have the potential to influence root growth. Species of *Azospirillum* commonly live in the rhizosphere of grasses. Some bacteria in the genus *Azospirillum* increase root growth by contributing nitrogen to the plant and others stimulate root growth by hormone production. Strains of species of *Azospirillum* vary in their capacities to fix atmospheric nitrogen or produce hormones. Nitrogen fixing strains of *Azospirillum* are not expected to increase root growth if nitrogen levels in the soil are already adequate for the particular plant species; if root growth is affected, other mechanisms must be involved.

Changes in root structure alter the capacity of root systems to absorb nutrients and water. Principally, this is related to the way that roots are distributed in soil. However, close associations between roots and microorganisms enhance nutrient uptake by plants if organisms have access to nutrients to which the plant does not. An example of this is the association between roots and nitrogen-fixing bacteria. Another example is the association between various types of mycorrhizal fungi and roots that allow root systems access to nutrients well beyond the rhizosphere.

Root pathogens commonly interfere with nutrient and water uptake by damaging the root surface and root cortex or by blocking the xylem vessels in the vascular system inside the root. Damage to the cortex or the vascular system will reduce nutrient and water transport between roots and shoots. Soil animals such as nematodes may also damage roots. Parasitic nematodes induce gall formation on some plants and animals such as springtails graze on small roots and damage their surface. Disturbances of this kind create opportunities for a variety of fungi and bacteria to enter roots that would not otherwise be able to do so.

Mycorrhizal Associations

The majority of plant species form mycorrhizal associations. Mycorrhiza means 'fungus root' and refers to several specialised associations between fungi and roots. There are four major types of mycorrhizas: ectomycorrhizas, arbuscular mycorrhizas (also called vesicular arbuscular mycorrhizas), ericoid mycorrhizas and orchid mycorrhizas. These associations are generally beneficial to both the plant and the fungus, with nutrient exchange between the partners. However, under some conditions, at least for arbuscular mycorrhizal, there is no beneficial effect and plant growth can even be reduced.

Orchid mycorrhizal associations are very different from the other three types. The minute orchid seed has no reserves of carbon and is totally dependent on the fungus during the early stages of its growth. Almost all plant species form one of the four types of mycorrhizal associations. Each plant species usually forms only one type of mycorrhizal association. In the less common situation in which more than one type of mycorrhiza is formed on the same plant, the different mycorrhizas usually occur on different roots or are present at different times during the life cycle of the plant.

One type of mycorrhiza is usually dominant in each ecosystem. For example, plants that form ericoid mycorrhizas often dominate heathland ecosystems. Trees that form ectomycorrhizal associations commonly dominate forests in Europe and North America. Tropical forests may contain more equal proportions of plant species that form arbuscular and ectomycorrhizal associations. The type of mycorrhiza present generally depends on the diversity of plant species that occur there. The eucalypt trees of Australian forests predominantly form ectomycorrhizal or arbuscular mycorrhizal associations. For example, in a jarrah forest in south-west Australia, seventy two per cent of the 93 species of plant studied formed either or both of these two mycorrhizas. The 25 per cent found with no mycorrhizal association were mostly members of the family Proteaceae, which does not form mycorrhizas but they can form specialised roots 'proteoid' roots which facilitate nutrient uptake.

Most annual agricultural plants form arbuscular mycorrhizas. Ectomycorrhizal associations are less common in disturbed ecosystems and are more frequently observed on perennial plants. Horticultural species form a variety of mycorrhizal associations, depending on whether they are ornamentals (all types of mycorrhizas are represented in this group) or vegetable and orchard plant species (most of these form arbuscular mycorrhizas).

Arbuscular mycorrhizas are the most common type of mycorrhiza and they are formed by a fungal group that occurs in most soils. The distinctive fungi that form arbuscular mycorrhizas have co-evolved with plants over millions of years. However, worldwide, there are relatively few species of arbuscular mycorrhizal fungi in comparison with other groups of soil organisms. Fewer than 200 species have been described, although many others occur which have not yet been formally named and described. These 200 species all occur within a relatively small number of genera including: *Glomus, Acaulospora, Archaeospora, Gigaspora* and *Scutellospora*. The genus of each fungus can be identified from the characteristics of their spores and molecular analysis is now also used to distinguish among species and genera. The most characteristic feature of arbuscular mycorrhizal fungi is the arbuscule. Arbuscules occur within cells of the plant and are key sites for nutrient exchange between the two partners of the association. The distinctive vesicles formed inside roots by some, but not all of these fungi contain high concentrations of lipids and appear to be storage structures.

Biology of Arbuscular Mycorrhizal Fungi

The roots of most plants are colonised by several species of arbuscular mycorrhizal fungi at the same time. There is little evidence of specificity between particular plants and these fungi, although some plants may be colonised to a greater extent by some of them. The relative abundance of different arbuscular mycorrhizal fungi within roots depends on soil conditions, root development and characteristics of the fungi such as the rapidity of spore germination and growth of hyphae.

Arbuscular mycorrhizal fungi are dependent on the plants they colonise for their survival. The fungi are "biotrophic" which means that they are unable to complete their live cycle without a plant. Because of this unusual characteristic of arbuscular mycorrhizal fungi, it is difficult to grow large quantities of arbuscular mycorrhizal fungi on artificial media in the laboratory without a plant. When newly formed hyphae of AM fungi intercept a root, the fungus enters with no apparent alteration to the root surface. Species, and even strains, of the same fungus colonise roots to different extents, but generally the more hyphae there are in the soil the greater the extent of mycorrhizal root formed. Arbuscular mycorrhizal fungi can survive for long periods in the soil as spores or as pieces of hyphae, even when the soil is dry or frozen. Most species produce spores that have relatively thick walls that are resistant to desiccation and are not easily eaten by organisms in the soil or colonised by pathogens. They have been shown to contain bacteria that appear to be natural inhabitants. Spores of arbuscular mycorrhizal fungi germinate when soil conditions become suitable, providing they are not in a dormant phase. Pieces of hyphae can survive in dead root pieces or directly in the soil.

Effects of Arbuscular Mycorrhizal Fungi on Plants

An important impact of AM fungi on plants is the alleviation of phosphate deficiency. Arbuscular mycorrhizal fungi form a network of hyphae in soil surrounding roots, and in doing so, effectively extend the root system. This can increase uptake of phosphorus from soil into roots. For example, some plants that are well colonised by arbuscular mycorrhizal fungi can grow just as well with only half the amount of phosphorus available in soil as that required by plants that do not have these fungi within their roots.

The extent to which a plant benefits from its arbuscular mycorrhizal fungi depends on the availability of nutrients and water. In terms of phosphorus nutrition, only plants growing where there is an inadequate supply of phosphorus will receive any benefit from having arbuscular mycorrhizas (this also requires that water and other nutrients are not limiting to growth). In soils where there is very little available phosphorus or where adequate phosphorus is available, there is generally no direct nutritional benefit of arbuscular mycorrhizas. Plant root architecture also influences the ability of a plant to supply itself with phosphorus, and therefore its dependence on an association with arbuscular mycorrhizal fungi. Plants such as grasses have fibrous root systems and are capable of thoroughly exploring soil for phosphorus. As a result, they do not benefit nutritionally from mycorrhizas to the same extent as a plant such as clover which has a coarser root system.

This effect of arbuscular mycorrhizal fungi on plants through increased phosphorus uptake has usually been studied only for young plants or plants grown in controlled conditions. The contribution of arbuscular mycorrhizal fungi to the growth of older plants, especially those growing in field soils, is not well understood. Obviously the relationship between the way the symbiosis functions and the nutritional status of the plant is complex and the plant does not always benefit. For example, under some conditions, arbuscular mycorrhizal fungi reduce plant growth, possibly because they take a large amount of carbon from the plant. Alternatively, *Scutellospora calospora* can increase plant growth and phosphate uptake into plants when the level of phosphorus in soil is extremely deficient. But, it can also decrease plant growth when the quantity of phosphorus in the soil is almost sufficient for the plant to reach its maximum potential growth. At higher levels of available soil phosphorus, the fungus appears to remove excessive carbon from the plant to an extent that is detrimental to the growth of the plant.

The abundance of arbuscular mycorrhizal fungi within plant roots changes the supply of phosphorus. Very low levels of phosphorus can be insufficient for the growth of the plant as well as the fungus, keeping fungal numbers low. As more phosphorus becomes available, both the plant and the fungus increase in growth. There can be an increase in the length of root colonised by arbuscular mycorrhizal fungi and a corresponding decrease in the proportion of the root colonised. This occurs if the fungus does not grow fast enough to keep up with the growth of the roots. In soil where phosphorus is present at an adequate level for maximum plant growth, the quantity of arbuscular mycorrhizal fungi in roots decreases substantially. This may either be due to a detrimental effect of excessive amounts of phosphorus on the fungi in the soil or to inadequate carbon supply in the roots for fungal growth.

Clearly, it is not always practical to classify a fungus as being either effective or ineffective at enhancing plant growth. The effectiveness of fungi can be altered by changing the amount of fungus in the soil, by changing the amount of phosphorus in the soil or by altering soil conditions that affect the growth of the fungus. In addition, strains of the same fungus may respond differently, so the diversity of fungi present is also significant.

Phosphorus is not the only nutrient taken up by the hyphae of arbuscular mycorrhizal fungi. For example, the supply of Cu (copper) and Zn (zinc) to some plants is improved in the presence of arbuscular mycorrhizal fungi. However, arbuscular mycorrhizal fungi do not increase the supply of nitrogen to plants. The reason that AM fungi enhance the uptake of some nutrients and not others is related to the way that nutrients interact with soil particles. Nutrients (such as nitrate) that move with water towards plant roots in most soils are not likely to become more available to a plant if it has arbuscular mycorrhizas. In contrast, P, Cu and Zn become adsorbed to soil particles and diffuse through a soil rather than move with water towards roots. Roots need to explore the soil to intercept these nutrients that move slowly by diffusion. By acting as an extension of the root system, the hyphal network of arbuscular mycorrhizal fungi increases the likelihood that roots have access to nutrients that diffuse slowly in soil.

Even if no nutrient benefit is gained from the presence of arbuscular mycorrhizal fungi, there can be a benefit to the physical state of soil if hyphae help to stabilise soil aggregates. Hyphae of arbuscular mycorrhizal fungi may also protect roots from disease. There is some evidence that arbuscular mycorrhizal fungi assist plants in water uptake, especially under dry conditions. Demonstration and interpretation of such benefits is confounded because the fungi also take up phosphorus. Enhanced growth may be due to increased phosphorus uptake as well as to water uptake if the soil is phosphate deficient. Therefore, increased water uptake may be an additional benefit in some circumstances, but the effects related to water and phosphorus are difficult to separate.

Within soil, arbuscular mycorrhizal fungi form extensive networks of hyphae that connect the roots of many plant species into an integrated system. This provides a pathway for nutrient exchange between different plant species. The extent to which this occurs, and its importance for plant growth needs further investigation. One particularly interesting area is the role of these networks in aiding seedling growth. Seedlings emerging from a soil containing a network of arbuscular mycorrhizal hyphae may gain access to nutrients from existing plants and grow better than might be expected if they are able to connect into this network. In addition, arbuscular mycorrhizal fungi may increase the chance of survival of seedlings in ecosystems where plants compete for nutrients and light. This is especially important for slower growing plants.

Therefore, successful establishment of arbuscular mycorrhizas may help maintain or increase plant species diversity in a natural ecosystem.

Ectomycorrhizal Fungi

Ectomycorrhizal fungi are the dominant mycorrhiza in many natural ecosystems, even though they form associations with a smaller number of plant species than do AM fungi. Plants that form ectomycorrhizas include well-known trees and many smaller perennial plants. The main features of an ectomycorrhizal association are the hyphal mantle on the root surface, the Hartig net formed between the cells of the root closest to the root surface and the extension of hyphae into the soil. The hyphae of ectomycorrhizal fungi do not enter root cells in the same way as AM fungi. Rather, the Hartig net is formed when the ectomycorrhizal fungi enter roots between epidermal cells and spread out to form a network of hyphae through the root. Ectomycorrhizal fungi are usually clearly visible on the surface of roots.

They promote root branching and restrict root extension which results in a great diversity in architecture of ectomycorrhizas. Roots become covered with hyphae to varying degrees. Ectomycorrhizas are formed by many different species of fungi, in contrast to the relatively small number of species that form arbuscular mycorrhizas. Most belong to families of the basidiomycetes and a number to families of the ascomycetes. Some are clearly visible in forests and other reasonably undisturbed ecosystems because they produce mushrooms and similar fungal spore-containing structures above the soil surface. Others form their reproductive structures below the soil surface. Small mammals dig up and eat these structures. Truffles are a well-known example of fungus-produced underground fruiting structures.

Some ectomycorrhizal fungi form associations with many different plant species, while others form associations with fewer plants. For example, most species of *Amanita* have an intermediate or broad host range, whereas most of species of *Suillus* have a narrow host range. In contrast, 25 per cent of species of Russula have a narrow host range and 30 per cent have a broad host range. Compared to AM fungi, many species of ectomycorrhizal fungi can grow in artificial, laboratory media, making them easier to study, although they often grow very slowly. Others are either difficult to grow or have not yet been grown at all under laboratory conditions.

Biology of Ectomycorrhizal Fungi

It is not easy to define a simple life cycle for ectomycorrhizal fungi because many genera of fungi are involved. Ectomycorrhizal fungi are commonly associated with perennial host plants that are relatively long-lived. Therefore, ectomycorrhizal fungi within the roots of living plants have the potential to survive for long periods.

Ecological successions of ectomycorrhizal fungi occur on the same root systems leading to a gradual change in dominance of fungi (Mason et al. 1984). Some plants form more than one type of ectomycorrhizal association, either on different roots of the same plant or at different times. In this study, fruiting structures of the fungus *Laccaria* sp. generally occurred further from the tree base than those of *Lactarius*. However, the presence of these structures is not a direct measure of the extent of mycorrhizal fungus present in soil or in roots. Recent molecular studies have highlighted the lack of a quantitative relationship between fungi in roots and the number of fruiting structures formed above the ground. New roots usually become colonised by the hyphae of ectomycorrhizal fungi attached to existing mycorrhizal roots on the same or on a nearby plant. The spores that are formed by these fungi do not appear to be very important for initiation of new mycorrhizas. Various degrees of specificity exist between ectomycorrhizal fungi and host plants. A process of molecular recognition takes place prior to colonisation of the root by the hyphae. Colonisation of the root by the fungus proceeds only if the fungus and root are compatible. Recognition of compatibility between a root and a fungus involves molecular signalling using genetic information in both the fungus and plant.

Effects of Ectomycorrhizal Fungi on Plants

It is more difficult to estimate the contribution of ectomycorrhizal fungi to plant growth than for AM fungi, except at early stages in plant development. Studies on seedlings clearly illustrate the contribution of ectomycorrhizal fungi in the early stages of tree development, but extrapolation of these benefits to older trees is not reliable. Ectomycorrhizal associations can improve the phosphate status of plants in soils that are phosphate deficient. Under some circumstances, ectomycorrhizal fungi make nitrogen more available to plants and increase plant resistance to drought and disease. These benefits can lead to a more robust plant that has a greater chance of survival. There is substantial evidence for benefits of ectomycorrhizal fungi in nitrogen nutrition of plants where nitrogen is deficient. Some ectomycorrhizal fungi can use simple or more complex forms of organic nitrogen. For example, all three fungi in this study were effective at using glutamine as a nitrogen source, but none could use histidine, which is a very complex form of nitrogen. In contrast, *Elaphomyces* was able to use nitrogen from arginine, but the other two fungi were not. These differences are likely to be important in soils with low levels of inorganic nitrogen and high levels of organic matter. The nitrogen that is taken up by the ectomycorrhizal fungi is passed onto their plant hosts, improving what would otherwise be relatively poor growing conditions for the plant. Ectomycorrhizal fungi also appear to directly help plants avoid disease and survive periods of drought. Where the hyphal mantle covers the root, it provides protection from pathogenic fungi, which can not penetrate the hyphal layer. In addition to this, due to the presence of the ectomycorrhizal fungi changes occur in root physiology that minimises the susceptibility of the root to infection by pathogenic fungi or bacteria. The hyphal networks around mycorrhizal roots enables water to be taken up from parts of the soil that the roots cannot reach. Protection from both disease and drought are contributions that ectomycorrhizal fungi may make to various degrees at different stages in the growth cycle of the host plant. The contributions depend on the type of fungus and the soil conditions.

Ericoid Mycorrhizas

Ericoid mycorrhizas are the specialised mycorrhizal associations formed with members of the plant families Ericaceae (Smith and Read 2008). These mycorrhizas have a different morphology from the mycorrhizas already mentioned and play an important role for their host plants (Perotto et al. 2002). Ericoid mycorrhizal fungi colonise the individual cells of very fine hair-like roots directly from the soil, forming coils of hyphae within the root cells.

The hyphal connections to the soil coalesce around the roots as a fine network. The hair roots colonised by ericoid mycorrhizal fungi are short lived. Therefore, ericoid fungi need to continuously re-establish the mycorrhizal associations as the roots grow. There appears to be a high degree of specificity between the fungi that form ericoid mycorrhizas and their host plants. Details of the nature of the recognition process between the fungus and plant are well understood.

Some ericoid mycorrhizal fungi grow in pure culture, and analysis of these cultures shows that ericoid fungi are very different from each other. This was demonstrated by comparing the electrophoretic pattern of enzymes that decompose pectin after they were separated in a gel. The migration of enzymes in the gel indicated differences in the molecular weight of pectic enzymes from some fungi. This is indicative of differences in genetic characteristics among the fungi.

There is evidence that ericoid mycorrhizal fungi have certain characteristics similar to wood degrading fungi that may allow them to degrade complex components of organic matter in the soil. In comparison to a selection of ectomycorrhizal fungi, two ericoid mycorrhizal fungi had greater potential to degrade lignin and soluble phenolic molecules. This conclusion was based on the response of the fungi in four biochemical tests that measured characteristics of fungi important in degradation of lignin and other complex plant molecules.

Effects of Ericoid Mycorrhizal Fungi on Host Plants

Ericoid plants occur in nutrient poor soils and the mycorrhizal associations that they form are likely to be important in providing them with nutrients from the organic matter that would otherwise remain unavailable to the plant. In particular, nitrogen appears to be released from organic matter by the activity of ericoid mycorrhizal fungi. There is also a possibility that organic phosphorus is mineralised from organic matter by these fungi. Although ericoid mycorrhizal fungi have characteristics that are very different from those of both AM and ectomycorrhizal fungi, all three types of mycorrhizas coexist in many natural ecosystems.

This illustrates how plant-microbial associations respond to the nutrient resources in a soil. The morphology and physiology of orchid mycorrhizas are very different from those of the other three types of mycorrhizas described above. Orchid seeds are highly dependent on fungi. During the early stages of seedling growth, the fungi supply carbon to the growing orchid seedling.

This is the reverse of what occurs in other mycorrhizas. Some orchid mycorrhizal fungi have been isolated and grown in culture for many years and many belong to the genus Rhizoctonia or to related genera.

NITROGEN-FIXING ASSOCIATIONS

Nitrogen gas (N_2) constitutes nearly 80 per cent of the atmosphere, but in this form nitrogen is not available to plants. Several groups of soil bacteria convert nitrogen gas into ammonium (NH_4^+), which is a form of nitrogen that can be used directly by plants. This transformation is catalysed by the bacterial enzyme nitrogenase. Very few bacteria have the capacity to activate the nitrogenase enzyme system, and those that do occur as free living bacteria in soil, as loose associations with root surfaces or within highly specialised, symbiotic associations with plants.

Bacteria in the genus *Azospirillum* live in a loose association with root surfaces of grasses and other plants. This contrasts with the highly specific associations that form between cyanobacteria and fungi in lichens. Highly specific nitrogen-fixing associations are also formed between some tree species and bacteria (actinomycetes) in the genus *Frankia*. The more primitive gymnosperms such as cycads also form characteristic associations with cyanobacteria. The most widely studied of all nitrogen fixing associations are formed between legumes and bacteria from the family Rhizobiaceae.

The biological transformation of N_2 gas to ammonium is equivalent to the industrial transformation of N_2 that requires excessive application of heat and pressure. This highlights the uniqueness of biological nitrogen fixation and its practical benefit. Nitrogen-fixing bacteria that live in loose associations with roots are called 'associative' nitrogen fixers. Associative nitrogen fixing bacteria grow on the surface of roots and may also colonise the outer layers of a root by entering between epidermal cells.

A common genus of associative nitrogen fixing bacteria is *Azospirillum*. Species of *Azospirillum* are not all equally effective at converting atmospheric nitrogen to ammonium. There is conflicting evidence about the extent to which nitrogen fixed by these bacteria is available to plants. Even for the species of *Azospirillum* that are very effective at fixing atmospheric nitrogen, the nitrogen fixed is not immediately available to the plant. It is not until the bacteria die and are mineralised that the nitrogen becomes available to the plants.

Nitrogen fixed by *Azospirillum* is not immediately available to plants because of the complex processes of ammonium transfer through the bacterial cell wall. Ammonium excess to the requirements of the bacterium can be transported to the outside surface of the cell wall. However, another process can transfer the ammonium back into the cell, slowing the transfer of ammonium to the plant. Some mutant strains of *Azospirillum* are effective at permanently transferring ammonium from the bacteria.

The movement of molecules into and out of bacteria and fungi are complex processes. In the case of *Azospirillum*, an understanding of these processes shows why nitrogen is not immediately available to plants from bacteria that fix nitrogen in the root zone, at least while the bacteria are alive. The benefits of associative nitrogen fixers have been estimated in agricultural soils. The quantity of nitrogen fixed varies greatly. In very nitrogen deficient soils, the quantities of nitrogen fixed are not usually sufficient to overcome nitrogen deficiency in agricultural plants.

Legumes and Nitrogen Fixing Bacteria Function

Legumes are important plants in both natural as well as managed ecosystems. The majority of legumes form symbiotic associations with nitrogen-fixing bacteria. As a consequence of these associations, distinctive structures (called nodules) form on the roots of legumes. The bacteria multiply within the nodule cells.

Many kilograms of nitrogen per hectare can be converted every year by agriculture plants including crop (*e.g.* soybean) and pasture (*e.g.* clover) plants because of their associations with nitrogen-fixing bacteria. These associations are of great importance as a substitute for nitrogen fertilizer. Nitrogen fixing symbioses are also common in natural ecosystems.

Bacteria that form associations with legumes live in the soil as part of the microbial community. However, they do not fix nitrogen directly in the soil. Nitrogen fixation only occurs when bacteria form a nodule on a legume root. When an appropriate legume is present, only one in a million bacteria are necessary to form a nodule. In forming a nodule, the bacterium undergoes a highly specific interaction with the plant, involving several stages and complex signals between the bacterium and plant.

Associations between particular bacteria and their legume host plants are usually highly specific. For example, the bacteria that fix nitrogen in association with species of clover cannot fix nitrogen in association with species of acacia. Nevertheless, the same bacteria can form associations with different clover species and occasionally different bacteria will infect the same plant. However, nodules generally contain only one strain of bacterium that multiplies to fill the central core of nodule cells.

Although there is considerable specificity among root nodule bacteria and legumes, it is interesting that the same bacteria form nitrogen-fixing associations with lupin (*Lupinus* spp.), a crop legume and serradella (*Ornithopus* spp.), a pasture legume. Because the architecture of both the roots and shoots of these two plants is very different, the bacteria which colonises both are evidently highly adaptable.

The shape and size of nodules varies considerably, with each determined by the plant, not the bacteria. In some plants, nodules can continue to grow throughout the life of the plant, as for example with clover. Where this occurs on perennial species, layers of new nodule tissue are added annually. In contrast, nodules formed on other legumes have a limit to their growth and are termed determinant nodules. Soybean is an example of a legume that forms determinate nodules.

The bacteria that form symbiotic associations with legumes differ in their ability to fix nitrogen. The effectiveness of these bacteria can also be influenced by soil conditions. Therefore, when planting agricultural legumes it is necessary to select the most effective bacteria that will do best in the environment. This is also true when planting tree species for revegetation and forestry projects. The bacteria that fix the most nitrogen when in association with a legume are not necessarily those that form the most nodules. Conversely, it is not uncommon to find nodules formed on agricultural plants by bacteria that are ineffective at fixing nitrogen. In soils where there are different strains of bacteria that can nodulate the same legume, many factors influence the success or failure of nodulation. Therefore, the full potential for nitrogen fixation is not always achieved, even if highly effective strains of bacteria occur in the soil.

Diversity Among Root Nodule Bacteria

There is considerable diversity among root nodule bacteria (McInnes et al. 2004). Bacteria that form nodules on legumes differ in the rate at which they multiply on artificial laboratory media.

Acacias form nodules with groups of bacteria with very different characteristics and this can affect the detection of bacteria on growth media if the rapidly growing bacteria overgrow those that grow more slowly.

There is a difference in the DNA structure among root nodule bacteria. In some, but not all, DNA is present as circular segments, separate from the main chromosome. These segments, called plasmids, contain the genetic information required to initiate symbiotic nitrogen fixation. Plasmids can replicate separately from the chromosome and transfer from one bacterial cell to another.

Genetic Basis for Nodulation and Nitrogen Fixation

The location of genes on the DNA that code for different stages in the nodulation process illustrates complex interactions that take place between legumes and root nodule bacteria during nodulation. Separate regions of bacterial DNA are responsible for recognition of the legume root surface, initiation of root hair curling, initiation of the infection thread formation, and root cell multiplication to form the nodule.

Bacteria present in the rhizosphere stimulate the plant to produce molecules called flavanoids which are released in root exudates. In turn, these molecules can, to varying degrees activate the nodulation genes in the bacteria. For successful nodulation to occur, the flavanoid molecules need to be present in sufficient quantities to induce a response in the bacteria. Some plants are better able to do this for some bacteria. In this example, the maximum activity of flavanoids produced by *Medicago murex* occurred at pH 6.0 whereas flavanoids produced by *Medicago littoralis* were most active at a higher pH.

This response partly explains why nodulation changes with soil pH for some legumes. Even though the bacteria may grow well at a certain soil pH, the lack of production of exudates that will initiate nodule formation may hinder their ability to form a nodule.

Legumes in natural ecosystems grow in soil containing indigenous, diverse communities of root nodule bacteria. Similarly, highly managed agricultural systems may contain many different strains of bacteria that have been introduced with agricultural legumes. High diversities of root nodule bacteria can result from a lengthy history of growing different legume species on the land, multiple introductions of commercial inoculant root nodule bacteria, and ongoing genetic diversification of bacteria in soil.

An analysis of the range of strains of root nodule bacteria that occur in Oregon pastures illustrates the level of genetic diversity that is possible. In this study, the bacteria were characterised using serological methods with four distinct serogroups identified and one most abundant in the surface 10 cm of the soil profile. Subsequent investigations using DNA analysis confirm the diversity of strains of rhizobia in soils. When bacteria selected for a particular legume are introduced into a soil, they have to survive as part of the community of soil bacteria. Otherwise, the only way they will be maintained in the soil is by regular re-introductions. The newly introduced bacteria also need to compete with existing strains of similar bacteria for nodulation sites on the legume root. For example, the strain of bacterium that was one of the original inoculants for soybean in the USA was subsequently found to be less effective at fixing nitrogen than some strains introduced later. The original bacterium was very competitive at forming nodules on soybean in comparison to the more recently identified effective strains. It is difficult for a newer bacterium to outcompete older bacteria that have already colonised the soil and are highly competitive at forming nodules.

This is clearly a major problem for the introduction of strains that are more effective at fixing nitrogen. A similar obstacle occurs when two clover species are grown in the same pasture. Nitrogen fixation occurs most effectively for each with different strains of *Rhizobium leguminosarum* biovar *trifolii*. However, it is not possible to ensure that each clover species is nodulated by the most appropriate strain of bacterium. The result could be that neither clover species fixes nitrogen according to its potential.

Bacteria that form legume symbioses can survive in a soil in the absence of suitable plants. Once the roots of a suitable plant reappear, the bacteria can re-initiate the symbiotic association. Some bacteria can survive without a plant for longer than others can. Even if nodule formation does not occur as soon as the plant reappears, the plant can stimulate the growth of the bacteria and make it more likely that nodulation takes place when other soil conditions are suitable.

Non-legume Plant Species form Nitrogen-fixing Associations

Frankia is a genus of bacteria (actinomycetes) that has the capacity to fix nitrogen gas in association with several tree species. Examples of such trees include the genera *Casuarina* and *Allocasuarina* in the family Casuarinaceae and *Alnus* in the family Betulaceae.

The associations formed by species of *Frankia* are generally host-plant specific, but the same bacteria can nodulate closely related plant species.*Frankia* associations with plants range from those that are highly effective to those that are relatively ineffective at fixing nitrogen.

Nitrogen deficiency in *Casuarina* can be remedied by inoculation with strains of *Frankia* that are effective at fixing nitrogen (Reddell and Bowen 1985). Two of the five sources of Frankia studied were effective at increasing growth of *C. equisitifolia*. Another was highly effective with *C. cunninghamiana*.

The association between *Frankia* and roots of their host plant is highly specific. The infection process is initiated through molecular signalling between the bacteria and root as for legume symbioses. When there is a compatible combination of bacteria and plant, molecular communication leads to the formation of a nodule on the plant root that develops a colony of bacteria within it. The bacteria stimulate the plant to form nodule tissue. Nodules can live for a long time, with a new layer of actively growing root cells and bacteria added each year. The bacteria fix nitrogen gas that diffuses through the walls of the nodule cells; the fixed nitrogen is transferred into the root cells and from there into the rest of the plant through the vascular system. A variety of other associations also occur between nitrogen-fixing organisms and plants. For example, *Macrozamia riedlei*, a member of the family Cycadaceae, forms an association with nitrogen fixing cyanobacteria. This plant forms highly branched roots that contain the cyanobacteria.

Soil-borne Plant Pathogens

Soils contain diverse communities of microscopic organisms that are capable of damaging plants. A detrimental interaction between a soil organism and a plant is often highly specific. For example, a fungus that causes root-rot of wheat may have no effect on the roots of another plant growing in the same soil. Highly specialised interactions between soil organisms and plants can kill seedlings and even adult trees.

Many organisms target younger plants but others appear as problems at later stages in the life of the plant.

Other pathogens are able to cause disease in many different plant species. The soil organisms that have the potential to be plant pathogens include fungi, bacteria, viruses, nematodes and protozoa. Some pathogens of the above ground parts of plants (leaves, stems) survive in the soil at various stages in their life cycles. Therefore, a soil phase of a plant pathogen may be important, even if the organism does not infect roots.

In spite of the potential for severe damage to be inflicted on plants by soil organisms, most plants do not display serious symptoms of disease. Disease usually occurs when conditions are particularly unfavourable, or when a soil organism is accidentally introduced where a highly susceptible plant species occurs. Intensive production in agriculture, horticulture or forestry increases the opportunities for diseases to develop compared with undisturbed natural ecosystems.

Planting of similar plant species together in monoculture increases the probability of disease outbreak. In contrast, the damage caused by the fungus *Phytophthora cinnamomi* in many different plant species in diverse natural ecosystems in Australia demonstrates the damage that can be caused by a pathogen that infects the roots of many unrelated plants.

Some plant pathogens depend on their host plant for survival and are unable to complete their life cycle without infecting their host plant.

Biotrophic organisms of this type are often difficult to grow in laboratory media. Disease-causing microorganisms and soil animals are a natural component of the soil community. The organisms are normally present in relatively low numbers. An outbreak of disease commonly follows either an increase in the abundance of the pathogen or a change in the susceptibility of the host to the pathogen.

Important questions are:

- Why do outbreaks of disease occur?
- What are the soil conditions that allow disease to develop or that suppress disease?
- What are the soil conditions that suppress disease?
- How do management practices influence the soil as a habitat for plant pathogens?
- How do plant pathogens survive in the absence of their specific host plant?

Plant Disease

Plant disease is the result of a complex interaction between the host plant, the pathogen and the environmental conditions including those in the soil. Soil conditions that influence the development of disease are pH, moisture, oxygen, and nutrients. In addition, other soil organisms interact with the pathogen. The level of this interaction is also influenced by soil conditions, which affects the growth and activity of the non-pathogenic organisms. Consequently, whenever the soil environment changes, there is potential for either a positive or negative influence on soil-borne plant pathogens.

Plant disease reduces plant survival and productivity, but the symptoms of plant disease are often difficult to recognise. One reason for this is that they are not necessarily the same in plants of different ages. The amount of a potential pathogen in a soil and the environmental conditions can also alter the symptoms displayed by a plant.

Furthermore, symptoms may be related to infection by more than one organism. Interactions between organisms can produce symptoms that are not the same as those that occur when each type of organism is present alone.

Similar symptoms can be caused by different pathogens and furthermore, the symptoms of some pathogens resemble those of various nutrient deficiencies. Therefore, it is not always easy to identify a pathogen responsible for a disease simply from the symptoms on the plant. A great deal of experience and knowledge of the range of symptoms produced by different pathogens and knowledge of symptoms of nutrient deficiency is essential for an accurate identification of most pathogens.

MITES

Although these are among the most common soil-dwelling mesofauna, they tend to be restricted to the leaf litter and surface layers. They are more diverse than any other single group of soil arthropods (including the insects), and this diversity is reflected in their feeding habits and life history.

In general, they are more important in their role as resource biota than as destructive biota.

Some species feed on fungal spores, while numerous predatory species attack nematodes, other mites, insect eggs, and larvae. Other species feed on plant debris, dung, or carrion and are important members of the decomposer community. Although some mite species feed on living plants and others are parasitic on livestock, in general these potential pests are only a minor component of the mite fauna.

Insects

The number of taxa of soil-dwelling insects is relatively small compared with the diversity that marks other major groups of soil organisms. With many species, only part of the life cycle (egg, larva and/or pupa) is spent in the soil, although some taxa are associated with specific soil types. Root-feeding species can reduce crop yields by reducing a plant's ability to absorb water and nutrients and may cause further losses by facilitating the entry of soil-borne pathogens. In contrast, predatory and parasitic insects can contribute to the biological control of both invertebrate pests and plant pathogens. *Some Collembola,* for example, show a marked preference for feeding on the spores of fungal pathogens, including *Rhizoctonia solani* and *Fusarium oxysporum.*

Species that feed on organic matter (detritivores) may be important members of the decomposer community. Some insects have ambivalent roles. For example, although termites are usually viewed as pests, in arid tropical soils with low earthworm populations the burrowing activities of termites can also help decompose organic matter and improve soil structure and porosity. Termites and ants are usually the dominant components of the soil insect biomass. As such, they are probably more important than all other insects in their effects on soil structure. They can also be significant pests. In parts of Malawi, for example, it has been estimated that the two most destructive termites (*Pseudacathotermes militaris and Macrotermes michaelseni*) damage crops on 72 per cent and 49 per cent of all smallholdings, respectively. In Africa as a whole, the subfamily *Macrotermitinae* is considered the most damaging group, affecting a wide range of crops, including tree and pasture species.

Their success as pests has been attributed to their ability to survive cultivation, feed on both living and dead plant material, and cultivate saprophytic fungi that serve as food in the dry season. Other important root-feeding insects include the 'white grub' larvae of various scarab beetles (*Scarabeidae*), weevils (*Curculionidae*), wireworms (*Elateridae*), and false wireworms (*Tenebrionidae*). Some *Hemiptera, Orthoptera,* and larval *Lepidoptera* (*e.g.*, cutworms) are likewise important in parts of the world. The significance of different species as pests is related to their host range. Those capable of feeding on a wide variety of plants, so called polyphagous species, are generally the most difficult to control. The distribution and damage potential of many soil-dwelling insects is also strongly influenced by soil moisture and rainfall patterns, with the more successful species being able to survive periodic drought. As noted earlier, damage by root-feeding insects can result in secondary infections by plant pathogens. Root crops are particularly susceptible to this type of damage, since the harvested part of the plant is directly affected.

Potatoes, for example, soon become infected with bacteria and fungi if damaged by larvae of the potato tuber moth (*Phthorimaea operculella*), resulting in the rapid rotting of the tuber. The same is true of sweet potato tubers damaged by larvae of the sweet potato weevil (*Cylas formicarius*). In groundnuts, the level of aflatoxin (caused by the fungus *Aspergillus flavus*) can increase if the pods are damaged by termites, since the latter can transmit fungal spores.

Bacteria

The most abundant members of the vast community of soil organisms are bacteria. Per gram of soil, they can reach densities of one billion (one thousand million) individuals—and an estimated 20 000 to 40 000 species. The magnitude of bacterial biodiversity has only recently been revealed through molecular techniques that can differentiate between hard-to-culture taxa. Bacteria play important roles in many soil processes, including the cycling of nitrogen, carbon, and phosphorus, and the degradation of pesticides and other potential pollutants. They can multiply rapidly under favourable conditions, although very high growth rates are generally limited by the availability of nutrients.

Bacterial populations are typically greater and more diverse close to plant roots. It has been estimated that 5 to 10 per cent of root surfaces may be occupied by bacteria. Various endophytic bacteria (*e.g.*, species of *Azotobacter*, *Acetobacter*, and *Azospirillium*) not only fix nitrogen but also stimulate the production of root hairs. This increases the host plant's capacity to take up water and nutrients and compensate to some extent for grazing by root-feeding pests. Various bacteria that colonise seed coats and plant roots produce compounds capable of affecting plant growth. These growth-promoting rhizobacteria (including species of *Pseudomonas, Bacillus, Serratia* and *Arthrobacter*) can also increase plant resistance to pests and diseases through a variety of mechanisms, including the production of structural materials that strengthen root tissues and antibiotic metabolites that directly affect other elements of the soil biota.

In some cases, these beneficial bacteria displace potential pathogens by competing for nutrients. Or, they may stimulate increased synthesis of defensive compounds by the plant itself, resulting in so-called induced resistance or systemic acquired resistance. Rhizobium etli, for example, induces systemic resistance in potato roots to the potato cyst nematode (*Globodera pallida*). Some rhizobacteria are able to suppress weed growth while leaving crops unaffected. The relationship between bacterial endophytes and their host plants is a subject of increasing interest to those seeking better ways to manage the soil biota. Plant roots themselves produce a number of compounds that affect microbial populations in their vicinity (the so-called rhizosphere). These compounds may act as attractants, repellents, biocides, or biostats (compounds that inhibit microbial growth), and as such offer new possibilities for the manipulation of both pathogens and beneficial species.

Fungi

As with the bacteria, the great diversity of fungi remains poorly documented. It has been estimated that only about 5 per cent of fungi have so far been described. Fungi play as important a role in soil processes as do bacteria, but tend to be more abundant in slightly acid soils. They vary widely in size, preferred habitat, and mode of life. Fungal plant pathogens (*e.g.*, some species of *Fusarium* and *Verticillium*) can cause diseases such as root rots and vascular wilts that are significant problems in many parts of the world. In the early 1990s, for example, the increasing prevalence of fungal root rots severely restricted the viability of bean crops in parts of Kenya. Many soil-borne fungal pathogens (*e.g.*, *Rhizoctonia solani* and *Pythium* spp.) are capable of infecting a range of plant genera.

Furthermore, plants infected with fungal diseases may be more vulnerable to attack by soil-dwelling insects. In Malawi, for example, it was found that groundnuts were more vulnerable to attack by termites if they were infected by fungal pathogens such as *F. solani*. However, some fungi form symbiotic relationships with plant roots, enhancing the plant's ability to take up nutrients. There is some evidence that such relationships can help less competitive plants to become established in pasture systems. Many species of soil fungi are saprophytic (*i.e.*, grow on dead organic matter), while others are parasitic on animals or plants. Some are important antagonists or biological control agents of soil-borne pests or diseases (*e.g.*, species of *Dactylaria* and *Arthobotrys* for nematodes, *Beauveria* and *Metarhizium* for insect pests, and *Trichoderma* and *Coniothyrium* for plantpathogenic fungi).

Nematodes

Nematodes (roundworms or eelworms) are the most abundant microfauna in the soil and are particularly numerous in the top 5 cm. In the top 2 cm of soil, their numbers may exceed two billion per hectare. The number of nematode species worldwide has been estimated at 80 000 to 100 000 species. The majority of soil-dwelling nematodes feed on bacteria and fungi, but many prey on other nematodes, protozoa and rotifers, and some species are parasitic on insect pests. More than 2000 species are parasites of higher plants. These include cyst and root-knot nematodes, such as certain species of *Heterodera, Globodera, Cactodera,* and *Meloidogyne*), and 'migratory' species such as *Pratylenchus* spp., *Ditylenchus destructor,* and *Scutellonema bradys.*

Their feeding lowers crop yields by disrupting water and nutrient uptake or by decreasing fruit or tuber quality or size; it can also allow fungal and bacterial pathogens to gain access to damaged roots, causing secondary infections. The presence of root-knot nematodes, for example, is known to increase the incidence of root rots and fusarium wilts in a wide range of crops. Plant parasitic nematodes tend to be more damaging pests in the tropics than in temperate zones. This is because their population growth rate is favoured by warm, humid climates and the longer growing seasons in these regions, which allows for more reproductive cycles per year. Nematode pests are associated with almost all crops grown in the tropics and cause losses of millions of dollars each year.

It has been estimated that, worldwide, plant-parasitic nematodes annually reduce agricultural production by about 12 per cent. Losses are particularly severe in developing countries due to insufficient expertise for species identification, inadequate quantification of the pest problem, and a limited range of management options. Worldwide, root-knot and cyst nematodes are especially important pests and infestations by them continue to undermine national efforts to improve food security and alleviate poverty.

Losses to individual crops often go unnoticed or are attributed to other causes. This is because most symptoms of nematode damage such as chlorosis (yellowing), patchy growth stunting, and wilting in hot weather, are easily confused with nutrient deficiencies or sometimes with bacterial or fungal diseases. Compared with the plant-parasitic nematodes, much less is known of the biology and ecology of other nematodes. This is particularly true of the bacteria-feeding nematodes, which are generally considered to be beneficial or harmless. These species, often concentrated in the root zone of higher plants, play an important role in soil nutrient cycling and can help distribute bacteria that promote plant growth. Although nematodes are not capable of widespread dispersal on their own, in agricultural systems they may be spread via contaminated machinery, land levelling, irrigation, and soil erosion.

Earthworms

The Greek philosopher Aristotle referred to earthworms as the "intestines of the earth". The description is apt not only because these invertebrates physically resemble digestive organs but also because they perform something of a digestive function when they break down plant residues in the soil and recycle nutrients. Experiments with maize, for example, have shown that plant residues can degrade 30 per cent faster in surface soil containing earthworms than in soil without them. Earthworms perform a variety of other ecosystem services as well. For instance, they enhance soil porosity thereby reducing rainwater run-off and allowing for infiltration of moisture to plant roots. They also play a vital role in water purification, agrochemical detoxification, and maintenance of soil stability.

When it comes to sheer biomass, earthworms usually account for the bulk of invertebrates that make their home in the soil. Along with ants and termites, they have a special distinction as 'ecological engineers'. This is due to their ability to excavate remarkably large amounts of soil and to create organomineral structures through their casts (excrement). To date about 3700 species of earthworms have been scientifically described. However, the actual number is estimated to be at least twice that, with the overall knowledge gap being significantly greater for tropical earthworm species than species in temperate zones. Although many aspects of earthworms' soil engineering role have been documented, interactions with other members of the soil biota, and how those affect plant growth and health, positively or negatively, are much less well understood.

Weeds

In the tropics, crop losses due to competition with weeds are on the order of 25 per cent, with weed removal often the most laborintensive task in smallholder systems. In most areas, the weed population centers on about 20 particularly troublesome species. The individual taxa may be indigenous or introduced, but they usually share a few key traits, especially rapid vegetative growth, high fecundity, and persistence in the soil seed bank. Weed species that resemble the crop in their early stages (*e.g.*, grass weeds in cereal systems) are particularly difficult to deal with when hand weeding is the main control technique. Besides competing with the crop for water and nutrients, weeds can also act as alternative hosts for pest nematodes and plant pathogens. And by boosting the humidity around the base of crop plants, weeds also increase the likelihood of infections by pathogens such as *R. solani* and *Sclerotium rolfsii*. Some weed species are parasitic on certain staple food crops, in some cases causing total crop failure.

The role of weeds in relation to soil-dwelling insect pests has not received a lot of scientific attention, although in some cases weeds are known to act as alternative hosts, maintaining the insect pest between crop cycles. Wild *Ipomoea* species, for example, can support the sweet potato weevil (*Cylas formicarius*) between crops, and in Zimbabwe, the groundnut plant hopper (*Hilda patruelis*) can survive the dry season on a variety of different weeds, allowing it to invade groundnuts as soon as they emerge. But weeds may also play a constructive role by supporting higher numbers of soil-dwelling predators that serve as pest control agents.

SOIL ORGANISMS: THE RECYCLERS AND SOIL BUILDERS

Soil organisms - many are microscopic workers spending lifetimes endlessly recycling dead plant materials and maintaining the fertility of the soil. In areas where their life cycles are not damaged soils naturally conserve themselves or actually build. These include uncut forests, soils under ponds and lakes, prairies and meadows with permanent plant cover and to some extent no till and mulched fields and gardens.

"We know more about the movement of celestial bodies than about the soil underfoot."

–Leonardo Da Vinci, circa 1500's

We've managed to figure out how to measure aspects of the soil's chemistry - so we think we know all about the soil. I know when I do a little test and find out my soil's pH is 8 I know a whole bunch. But, equally, if not more important, are the soil organisms living within the mineral stew. The problem with soil organisms is that most of them are invisible to us. The most important, the bacteria and fungi, are microscopic and we think that they might be a problem. I mean have you ever considered loving your bacteria - or your fungi. Plus they just aren't all that cute. They're blind - which makes absolute sense given that they live in the dark - so they don't have eyes. They aren't pretty colours - most variations on the black and white theme. Basically you have to be someone like Grissom (from CSI) to love them.

Soil Organisms

Even the smallest amount of information about soil organisms can threaten your world view - and with luck entice you to love and care for these small beings. Or at least it has done so for me. I'm a committed organic type gardener and have been for some time, but, I have to admit I didn't really understand what was going on in that soil underfoot my feet touch everyday. I still don't really - it's too complex and brilliant for me to totally fathom. But thanks to a few soil nerds who've been kind enough to share their work with the world in books I've been partly enlightened. Here is the information in a nutshell.

Soil, or at least healthy soil, is teeming with life. This life lives in an extremely symbiotic and mutualistic way unfailing offering their services to help the whole system work.

In a thimble full of soil - about a gram weight - you can expect to find

- 100 million to 1 billion bacteria
- Several yards to several miles of fungi depending on whether you have agricultural or forest soil
- Several thousand protozoa and up to several hundred thousand in forest soils
- From ten to several hundred nematodes

This is all in a thimble full of soil. And this doesn't begin to count the larger mites, spiders and earthworms who together form a complex called the Soil Food Web. I suspect that we would love soil organisms a lot more if we could see them. But the fact is we can't see most of them and to a large extent out of sight is out of mind. Remember, in a natural system you don't need to be adding a bunch of fertilizers, pesticides, herbicides and fungicides every year to keep things growing and healthy. Natural systems maintain themselves thank you very much and they do this with the work of the soil food web.

First the Bacteria and Fungi - foundations of the web

- Hold nutrients in the root zone for your plants
- Build soil structure so that water both drains and holds better
- helps prevent soil erosion
- Produces plant stimulating hormones
- Fixes nitrogen from nitrogen gas to plant usable nitrogen
- makes phosphorus available in a plant usable form
- Open the soil so that plant roots can penetrate more deeply
- Compete with bad guy bacteria and fungi thus keeping them in check and helping plants fight disease or stay disease free

The other soil organisms, the protozoa, nematodes and other larger members of the soil food web eat the bacteria and fungi and in doing so release the nutrients the plants need when they need them. Pretty amazing really.

Care and Feeding of Soil Organisms

For a moment think of your soil organisms as very cute and lovable puppies. Naturally you want to give them a comfortable and safe home, good food, and the air and water they need. Guess what - soil organisms have similar needs. Unfortunately they aren't cuddly and you can't play with them, for the most part you can't even see them, so you might not notice that you're killing them with either your gardening or farming methods or sheer neglect.

Habitat - most of the soil creatures live in the top few inches of the soil. When we turn the soil over with a rototiller or a shovel we can bury our babies in a place where they no longer can function. Fungi have the worst time with our cultivating methods. Their filaments get chopped up and the long straws they were forming to syphon nutrients to their plant partners get severed and big parts cease to function. Many farmers have adopted no till methods in their fields to improve soils and this will serve to give the soil food web a chance to recover.

Air - when we walk on our gardens we compact the soil under our feet. This is many times worse when we drive over land with a tractor. What it does is eliminate some of the pores that were holding oxygen - always handy for the beneficial soil organism. As oxygen goes down the anaerobic bacteria and fungi go up and the bad news is these are the guys that cause plant diseases.

Feed Good Food - like us the soil organisms need good food. Bacteria tend to like the green foods and sugars and fungi the brown foods like woods and tough leaves. Compost, being a blend of green and brown foods, is the perfect solution - and it has the added benefit of increasing the numbers of bacteria, fungi, protozoa and nematodes at the same time.

Avoid Bad Food - In the past - and even now - we tend to feed the garden chemical fertilizer. This changes the habitat fast - too fast for our soil pets to move or adapt so they can die. Essentially we then put our plants on life support - something akin to an IV drip instead of real food. In doing so we kill the chance of real food coming their way through the network of the soil food web. Plus we of course have bills to pay - fertilizer bills.

Safety - here is where the -cides come in. When the biology of the soil gets out of whack we tend to have a bunch of problems crop up - insect pests run amuck, fungi and bacteria shift from help mode to destroy mode and weeds take over. So naturally we reach for pesticides, herbicides and fungicides. They are effective killers but like big bombs they tend to kill a lot more than we intended. We lose our soil food web and all it's benefits once again.

IMPORTANT SOIL ORGANISMS

Each trophic level of the soil food web is comprised of several types of organisms. This section will describe some of the most prevalent and important soil organisms; however, it is important to note that there are many other types of organisms that exist in the soil.

Earthworms

Earthworms are very important soil inhabitants. While not required for a soil to be healthy, they are an indicator of soil quality. The burrowing and ingestion of soil materials by earthworms helps to physically mix the soil, decompose plant materials, and cycle nutrients. The casts of earthworms can also improve soil structure by enhancing aggregation. Earthworm activity is usually low in Florida's sandy soils.

Arthropods

Arthropods are animals with an exoskeleton, such as insects, crustaceans, and arachnids. In the soil, arthropods are responsible for physically shredding organic materials, stimulating microbial activity, and cycling nutrients. Soil arthropods are categorised based on their function in the soil. Large arthropods like millipedes and termites are shredders that chew up dead plant materials. Herbivores like mole crickets feed on plant roots. Shredders and herbivores can become pests when their populations are high. Predator arthropods, including centipedes and spiders, help to control the populations of other soil arthropods. Some arthropods eat soil fungi and bacteria.

Nematodes

Nematodes are non-segmented worms, most that live in soil are microscopic. Many nematodes feed on bacteria, fungi, algae, or are predators of microscopic animals. Other nematodes are parasites of animals or plants. In the soil, nematodes assist with nutrient cycling, are a source of food for other soil organisms, and can suppress or cause disease. While root-feeding nematodes can cause plant disease, fungal-feeding, bacterial-feeding, and predatory nematodes often work to suppress plant diseases or pests.

Soil Protozoa

Protozoa are single-celled, mobile organisms that are important for nutrient cycling and regulating populations of bacteria in the soil. Protozoa, which are larger than bacteria, are often abundant in soils. Protozoa populations can reach up to 1 million per teaspoon of soil in highly fertile soils. When they feed on bacteria or other protozoa, these organisms often release excess nitrogen in a form that can be used by plants. Protozoa depend on soil moisture to feed and for movement. Therefore, soil moisture influences the type and size of protozoa in the soil. Soil protozoa are often most active near plant roots.

Bacteria

Bacteria are microscopic, single-celled organisms that are abundant in soils. In fact, a teaspoon of soil can contain up to 20,000 different species of bacteria. In the soil, bacteria are important for nutrient cycling, disease suppression, and even pollutant cleanup. Soil bacteria can also improve soil structure by enhancing the formation of aggregates. Similar to soil fungi, soil bacteria are categorised as decomposers, mutualists and pathogens. There is also a fourth type of bacteria called chemoautotrophs. Decomposer species are responsible for the breakdown of organic residues. Some mutualist bacteria species are capable of converting nitrogen in the atmosphere to a form that is available to plants. These nitrogen-fixing bacteria colonise the roots of plants called legumes (*e.g.*, clover, soybean, alfalfa). Chemoautotrophs obtain energy from nitrogen, sulfur, or iron compounds rather than from carbon. These bacteria are often capable of decomposing harmful chemicals like oil and pesticides that find their way into the soil.

Factors Affecting Soil Organism Populations

The type and number of organisms present the soil environment is directly related to the availability of food. However, other factors like climate, vegetation, and soil pH also impact the type and quantity of organisms present. For example, forests usually have more diverse soil organism populations than disturbed areas (*e.g.*, cultivated fields, urban areas). Highly disturbed soils, such as those found in urban areas, typically have smaller numbers of soil organisms than forest or grassland soils. Also, soil organisms are usually more active in areas with a warm, moist climate than in areas with cold or dry climates. Typically, soil microorganisms (*e.g.*, bacteria and fungi) account for most of the biological activity in soils.

4

Soil Structure

Soil structure is evaluated by determining one or more of the following parameters: (1) size of soil particles using dry or wet sieving methods, (2) pore-size and its distribution in soil, (3) saturated hydraulic conductivity of soil in situ, (4) bulk density of soil in situ. (5) infiltration rates of soil in situ, (6) hydraulic conductivity at compacted minimum bulk density, and (7) energy required to cause aggregate disintegration.

One of the methods followed in the Division of Agricultural Physics, Indian Agricultural Research Institute, for determining soil structure is the aggregate analysis by wet sieving method using a modification of the original Yoder-type sieve-shaker having four nests each containing sieves with apertures ranging from 2.0, 1.0, 0.5, 0.25 and 0.10 mm.

Air-dry soil samples (say 50 g each) having aggregates from 5 to 8 mm size are spread on the mesh of top sieve (2 mm) and are saturated with water by gradually increasing the level of water in the tank of the sieve--shaker. Samples are then sieved in water for 30 minutes by shaking at the rate of 33 cycles per minute.

The water stable aggregates retained on each sieve are then quantitatively transferred to tared aluminium cans, oven -dried and weighed. Oven-dry soil aggregates of various sizes of each sample are transferred to a container having 10 ml of 5% solution of sodium hexametaphosphate with sufficient distilled water to disperse soil particles.

After stirring the suspension for 15 minutes, the contents are washed in a stream of water on the same sets of sieves which were used earlier. The primary particles retained on each sieve are now oven- dried and weighed.

Moisture content of soil samples on oven-dry weight is also calculated separately to determine the percentage of aggregates. The calculation is as follows:

Percentage of aggregates > 0.25 mm =

$$\frac{\text{Weight of soil particles} > 0.25\text{mm} - \text{Weigh to sand} > 0.25\text{mm}}{\text{Oven dry weight of soil sample}}$$

Based on the above data, aggregation index can be calculated by deter-mining the area between the two curves obtained by plotting the accumulative percentages (fraction less than a particular size) of soil particles and sand against the sieve openings.

INTRODUCTION

Movement of air and water in soil is governed by the structure of soil. The structure of soil, in turn, is dependent on stable aggregates of soil particles. A soil aggregate is a naturalIy occurring cluster of soil particles with a strong binding force among them. Soil organic matter including humus, polysaccharides and polyuronides produced by soil microorganisms help to cement soil particles together while filamentous fungi provide additional mechanical support.

Several workers have found that the aggregating influence of different groups of soil microorganisms can be graded in the order, fungi > strep tomyces > gum producing bacteria > yeasts. Rapidly growing Rhizopus, Mucor, Chaetomium, Fusarium, Cladosporium, Aspergillus and Rhizoctonia are good examples of fungi known to secrete gums and provide mechanical support to bind soil particles. Examples of bacteria producing significant amounts of gum are Azotobacter, Beijerinckia, Rhizobium, Xanthomonas and Bacillus.

The lipopolysaccharides from gums of Rhizobium and Agrobacterium have been analysed and found to contain glucose, rhamnose, mannose, glucosamine and 4-0-methylglucoronic acid as major constituents. Other components of the gums are fructose, arabinose, galactose and xylose. It has been claimed that infection of root hairs of legumes is dependent on a water soluble polysaccharide of the rhizobial gum, although no conclusive evidence has been provided to support such a hypothesis.

Different soil components have been chemically extracted from soil and their effect on soil and their effects on soil aggregation properties studied. Among the carbohydrates isolated from soil, dextran containing high amounts of uronic acid which is resistant to microbial degradation has been shown to have the best soil aggregating qualities.

Several experiments have been carried out under laboratory conditions to demonstrate the effect of single or mixed cultures of microorganisms in soil amended with various carbon sources. The results of such experiments, however fascinating they may be, must be viewed with caution. It is also interesting to note that the involvement of polysaccharides in soil aggregation has been questioned by some workers based on results of experiments in which native polysac-charides of soil samples were destroyed by chemical treatments without any detriment to the property of soil aggregation.

ROLE OF ORGANIC MATTER

Soils receiving well decomposed organic manures have better soil aggregates than those receiving saw-dust and other types of not so easily decomposable organic wastes. The root hairs and exudates from roots together with sloughed off debris of the root cortex help in creating soil aggregates in the root zone which can be demonstrated by lifting a her-baceous plant carefully from soil.

Since organic matter decomposition involves microorganisms, the products of microbial growth such as gums rather than the quantity of organic matter added is the deciding factor in the improvement of soil structure. The complex polysaccharides of microbial origin are resistant to microbial attack.

This may be due to the ability of gums to form a complex with metals, other organic compounds and clay minerals in soil or due to physical protection offered by the aggregates preventing further microbiological degradation. Adequate evidence has not been forthcoming on these aspects to explain the persistence of microbially synthesised polysaccharides in soil and further research is needed on this aspect, under natural soil conditions.

SYNTHETIC SOIL CONDITIONERS

These are synthetic compounds which act as stabilizers of naturally occur-ring soil aggregates. Their value lies in their inherent ability to resist microbial degradation and not in merely helping in the formation of new aggregates.

They simulate microbial polysaccharides in binding soil aggregates and are known as Krillium type conditioners Polyacrylic acid (PAA), Polyacrylonitrile (PAN), hydrolysed Polyacrylonitrile (HPAN) and Vinyl acetate-maleic acid copolymer (VAMA).

Such compounds are active in concentrations of about 0.1-0.2% (per unit dry weight of soil) in an aqueous phase. Soil conditioners have not been very popular because of the high cost and the necessity of frequent application to obtain desired results.

RELATION TO ROOT GROWTH

Although soil structure is a complex of several physical characteristics of soil, the growth pattern and extent of root system together with the atten-dant exudates and sloughed off debris of the root cortex contribute additional factors to the maintenance of soil structure, especially under intensive cultivation as in multiple cropping, mixed cropping and relay cropping. In other words, the root system of an actively growing plant or the remains of root system after harvest contribute to the organic matter status of soil and hence indirectly influence soil aggregation. In this con-text, leguminous crops are known to improve soil structure.

The growth of a plant depends not only on the capacity of soil to release nutrients, but equally on the capacity of the root system to absorb such nutrients in an efficient manner. An ideal root system is one which has the following characteristics: (1) capacity to spread quickly into a large volume of soil, (2) ability to spread spatially to hold the shoot in an erect position, (3) capacity to penetrate soil aggregates and soil layers of varied compaction and absorb water and nutrients from subsoil zones, (4) capacity to solubilize and absorb maximum nutrients and moisture, especially under drought conditions, and (5) add large quantities of organic matter to different soil layers to improve physical properties of soil. Varieties of one and the same crop show differences in their ability to add organic material through roots into the soil and thereby improve soil structure. For instance, in subsoil depths, Kalyan Sona variety of wheat could add 170 kg/ha whereas Sharbati Sonora could add only 60 kg/ha. Root systems of different plants also differ in their abilities to absorb native soil nutrients and nutrients from inorganic fertilizers.

This as verified from experiments in which 32p tracer technique was used to test the feeding efficiency of roots of different varieties of wheat and rice. In wheat, relative per cent phosphorus feeding efficiency of Sonalika variety was 72.4 for soil phosphorus whereas it was 27.6 for fertilizer phosphorus. Similarly, the per cent feeding efficiency of soil phosphorus versus fertilizer phosphorus in NP 130 variety of rice were 76.3 and 23.7 respectively. Invariably, the feeding efficiencies of native nutrients were also uniformly better than nutrients from fertilizers, in several other varieties of wheat and rice.

SOIL THERMAL PROPERTIES

The thermal properties of soil are a component of soil physics that has found important uses in engineering, climatology and agriculture. These properties influence how energy is partitioned in the soil profile. While related to soil temperature, it is more accurately associated with the transfer of heat throughout the soil, by radiation, conduction and convection.

The main soil thermal properties are:

- Volumetric heat capacity, SI Units: $J.m^{-3'''}K^{-1}$
- Thermal conductivity, SI Units: $W.m^{-1'''}K^{-1}$
- Thermal diffusivity, SI Units: $m^{2'''}s^{-1}$

Measurement

It is hard to say something general about the soil thermal properties at a certain location because these are in a constant state of flux from diurnal and seasonal variations. Apart from the basic soil composition, which is constant at one location, soil thermal properties are strongly influenced by the soil volumetric water content, volume fraction of solids and volume fraction of air. Air is a poor thermal conductor and reduces the effectiveness of the solid and liquid phases to conduct heat. While the solid phase has the highest conductivity it is the variability of soil moisture that largely determines thermal conductivity. As such soil moisture properties and soil thermal properties are very closely linked and are often measured and reported together. Temperature variations are most extreme at the surface of the soil and these variations are transferred to sub surface layers but at reduced rates as depth increases. Additionally there is a time delay as to when maximum and minimum temperatures are achieved at increasing soil depth (sometimes referred to as thermal lag).

Single and Dual Heat Probes

The single probe method employs a heat source inserted into the soil whereby heat energy is applied continuously at a given rate. The thermal properties of the soil can be determined by analysing the temperature response adjacent to the heat source via a thermal sensor. This method reflects the rate at which heat is conducted away from the probe. The limitation of this device is that it measures thermal conductivity only. Applicable standards are: IEEE Guide for Soil Thermal Resistivity Measurements (IEEE Standard 442-1981) as well as with ASTM D 5334-08 Standard Test Method for Determination of Thermal Conductivity of Soil and Soft Rock by Thermal Needle Probe Procedure.

After further research the dual-probe heat-pulse technique was developed. It consists of two parallel needle probes separated by a distance (r). One probe contains a heater and the other a temperature sensor. The dual probe device is inserted into the soil and a heat pulse is applied and the temperature sensor records the response as a function of time. That is, a heat pulse is sent from the probe across the soil (r) to the sensor. The great benefit of this device is that it measures both thermal diffusivity and volumetric heat capacity. From this, thermal conductivity can be calculated meaning the dual probe can determine all the main soil thermal properties. Potential drawbacks of the heat-pulse technique have been noted. This includes the small measuring volume of soil as well as measurements being sensitive to probe to soil contact and sensor to heater spacing.

Remote Sensing

Remote sensing from satellites, aircraft has greatly enhanced how the variation in soil thermal properties can be identified and utilized to benefit many aspects of human endeavour. While remote sensing of reflected light from surfaces does indicate thermal response of the topmost layers of soil (a few molecular layers thick), it is thermal infrared wavelength that provides energy variations extending to varying shallow depths below the ground surface which is of most interest. A thermal sensor can detect variations to heat transfers into and out of near surface layers because of external heating by the thermal processes of conduction, convection, and radiation. Microwave remote sensing from satellites has also proven useful as it has an advantage over TIR of not being effected by cloud cover.

The various methods of measuring soil thermal properties have been utilized to assist in diverse fields such as; the expansion and contraction of construction materials especially in freezing soils, longevity and efficiency of gas pipes or electrical cables buried in the ground, energy conservation schemes, in agriculture for timing of planting to ensure optimum seedling emergence and crop growth, measuring greenhouse gas emissions as heat effects the liberation of carbon dioxide from soil. Soil thermal properties are also becoming important in areas of environmental science such as determining water movement in radioactive waste and in locating buried land mines.

SOIL IN GENERAL

Soil is the outer covering of the earth which consists of loosely arranged layers of materials composed of inorganic and organic constituents in different stages of organization. It is the natural medium in which plants live, multiple and die and thus providing a perennial source of organic matter which could e recycled for plant nutrition.

Soil provides the physical support needed for the anchorage of the root system and also serves as the reservoir of air, water and nutrients which are so essential for plant growth. The portion of the earth beneath the soil in known as the bed rock and does not contribute directly to the growth of plants.

The processes involved in the formation of soil are slow, gradual and continous and are the sum total of environmental effects on rocks collectively known as the weathering of rocks. Rocks are exposed to sun, wind and rain. The changes brought about by fluctuations in temperature cause cracks on the surface of rocks in which the minerals released by the action of rain-water accumulate. The winds bring in particles of organic matter and sooner or later lichens establish on the rocks. The rocks gradually disintegrate resulting in a loose assemblage of rock materials which finally break down into smaller pieces of rocks.

The process of weathering starts afresh in the smaller rocks and continues as a chain reaction. What has now been described in few words in a process which normally takes place in nature involving hundreds or thousands of years, depending on the nature of the parent rock material.

In fact, weathering of rocks is a continuous phenomenon and will serve to add more and more soil to the surface phenomenon and will serve to add more and more soil to the surface of the earth.

There are different types of parent materials of rocks available for the formation o soil. Granite is slow to weather unlike limestone, sandstone or shale. The nature and chemical composition of the parent material have a direct berating on the fertility of soil. For instance, soils derived from granite or sandstone are no as fertile as those derived from limestone.

Broadly, speaking, there are six soil zones in the world the tundra soils having dwarf shrubs and mosses in very cold regions of the world, the podzolic soils of humid temperate forests, the chemozemic soils of subhumid, semi-arid temperature climates, the desertic (arid) soils of arid, temperate and tropical climates, the latsolic soils of humid and wet or dry tropical and subtropical climates and the stony mountain soils which may have any one of the above characteristics. Each soil zone or belt may consist of many soil types which can be clubbed into various soil groups depending on fertility, tilth, ability to hold water and resistance to erosion.

In India, approximately 25 soil types are recognized. Among them, those which have been extensively studied are the red soils, the laterite soils, the black soils and the alluvial soils. Besides these, types nearly 14% of the total area of soil in India is occupied by forest trees and hills and certain parts of India are characterized by desert conditions and alkalinity which render them unfit for cultivation.

Each type of soil is characterized by the presence of different horizons which can be seen when a soil profile is obtained. The formation of soil horizons (mostly distinguished as A, B, and C horizons) depends on climate, living organisms, parent rock material, topography and time, all of which control the weathering of rocks. All the three horizons and their subdivisions (often designated as A1, A2, A3, and so on) may not be present in all soils but a given soil is bound to possess at least some of them.

The A horizon is formed and differentiated by the accumulation of organic matter and its decomposition products through microbial activities. Transitional stages exist between individual horizons and. In practice, it is difficult to clearly demarcate such transitional areas. Nevertheless, the upper most horizon is rich in organic nutrients.

The B horizon is characterized by the accumulation of silicate clay minerals and percolating humans particles while the horizon C consists of weathered parent material. The A, B, and C horizons are interlinked physically, biologically and chemically whereas the lowermost D horizon may consist of hare rock, layers of clay or and which are not part of parent material but may have some relationship to the upper horizons of soil.

PHYSICAL PROPERITIES OF SOIL

Physical properties of a soil type depend on the size of particles in it. Particles above 2.0 mm are generally classified as gravel or stones and others as sand (between 0.05 and 2.0 mm), silt (0.002 to 0.05 mm) and clay (less than 0.002 mm). The texture of a soil depends on the percentage of sand, silt and clay in it. Based on this yardstick, soils are designated as clay, sandy clay, clay loam, silty clay, silty clay loam, sandy loam, silty loam, loamy sand, sand and silt. Soil particles occupy roughly more than half the space in soil. The remaining space between the particles, called the pore space, is occupied by water and air.

The bulk density of a soil is defined as the mass of a unit volume of dry soil which also reflects the total pore space of a soil. The pore size or porosity of soils together with bulk density determine the structure of soil. Sandy soils have single grained structured because of the general uniformity in particle size whereas in sandy clay or silty clay soils, the particles are grouped into aggregates.

The stability of soil aggregates depends on the organic matter content of individual soils and the nature of the microbial products which bind the particles together. Soil aggregation is a vital factor in plant growth, since movement of air, water and transfer of energy are interlinked with the porosity of soil.

Soil temperature is a very variable factor but nevertheless influences plant growth depending on the intensity of sunlight, day length, seasonal variations, rainfall and colour and texture of soil. Soil crusting is an important physical feature os soil which influences the emergence of seedlings. Soil crusts are formed by desiccation and may cause injury to the stems of seedlings resulting in crop failures.

The soil triangle showing the relationship between contents of clay, silt and sand in determining the different kinds of soil

Other physical factors in soil affecting plant growth are pH and moisture holding capacity. Retention of water in soils is related to pore space and capillary action of soil particles while pH is dependent on the chemical condition of soil.

Acid or alkaline soils are not generally as suitable for plant growth as neutral soils, since solubility and availability of plant nutrients are related to soil pH. Acid soils are usually characterized by excess availability of aluminium, iron, manganese, copper and zinc which may even prove toxic to plants.

The reverse is generally true of alkaline soils and in such soils plants show symptoms of deficiency to many of the elements listed above. Neutral soils, in particular, favour the growth of such microorganisms which are responsible for the conversion of organic forms of nitrogen, phosphorus and sulphur into inorganic forms which can be absorbed by plants.

Changes in temperature affect not only the physiological reaction of microbial cells but also the characteristics of soil environment. A sudden fall in growth rate of microorganisms can be the result of thermal denaturation of proteins and changes in the permeability of membrane at high as well as low temperatures.

However, specific proteins of flagella and ribosomes of themophilic bacteria are more thermostable than those of mesophilic bacteria which grow at ambient temperatures.

At low temperatures, microbes alter their cell structure especially the shape of proteins because cold hardiness of genes in ribosomes control the genetic coding of proteins and hence an organism's ability to function at low temperatures. Growth of most psychrophilic bacteria stops at zero degree temperature despite the fact that some species are capable of growing below the freezing point because of a mechanism to keep the cytoplasm unfrozen.

CHEMICAL PROPERTIES OF SOIL

Soil is the medium from which plants normally derive their nutrients. The nutrients are carbon (C), hydrogen (H), oxygen (O), nitrogen (N), phosphorus (P), sulphur (S), potassium (K), calcium (Ca), magnesium (Mg), iron (Fe), manganese (Mn), zinc (Zn), copper (Cu), molypdenum (Mo), boron (B), and chlorine (Cl). Off these sixteen elements, Fe, Cu, Mo, B and Cl are considered as micronutrients since they are required in trace amounts for plan growth and the remaining ones are classed as macronutrients because they are required in large quantities.

The three main components of the soil which provide nutrients for plants growth are the organic matter, the derivatives of parent rock materials and the clay fraction. The nutrients are first released into the soil solution (soil water) before they get transferred into the root system of plants. The organic matter in soils is a potential source of N, P and S for plant growth. Microbiological decomposition of organic matter is an essential step to release the bound nutrients in organic residues into an available form.

The inorganic minerals come from rocks which are made available to plants by mineral decomposition. On the other hand, the clay fraction of the soil provides secondary minerals and amorphous materials which are different form those derived fro sand and silt. They are clay minerals and hydrous from those derived from sand and silt. They are clay minerals and hydrous oxides.

The clay minerals are composed of three mineralogical types kaolinite, Montmorillonite and illite. The hydrous oxides are compounds of iron and aluminium. Kaolinite has a crystalline structure composed of two-layer uni cells, one layer of which is made of aluminium and oxygen octahedral and the other layer made of silicon and oxygen tetrahedra (four-sided) with silica and aluminium in a 1: 1 ratio.

The two layers are help together by common oxygen-hydroxyl linkage into a rigid structure that does not expand when wet and hence prevent the exchange of available cations. Montmorillonite has a units comprising of one layer of silicon and oxygen tetrahedra, another layer of aluminium and oxygen tetrahedra and one more layer of silicon and oxygen and hydroxyl bonds. Together, the silica and alumina are in a 2: 1 ratio.

Between the two units lies a space that expands when wet accommodating cations in an exchangeable dorm that are available for absorption of plants. Illite is similar in structure to Montmorillonite and does not expand when wet but has the capacity to fix potassium (K+) between nonexpanding plates.

The colloidal clay in soil acts as a large anion and absorbs cations. Having absorbed the cations, the clay particles act as reservoirs of exchangeable ions and release them for plant nutrition as and when required. Thus, the cation exchange or the base exchange properties of soil determine soil fertility and hence plant nutrition. The exchangeable cations are hydrogen, calcium, magnesium, potassium, sodium, ammonium, manganese, zinc, copper and aluminium.

These ions possess different energies by which they are held to the solid phase of soil. The binding energy of ions also determines the ease with which the ions may be exchanged in soil. In fact, the property of base exchange in soils enables ions to be held in a readily available state for plant nutrition and at the same time prevents the loss of nutrients by leaching.

There are both positively and negatively charged ions in the soil solution. The positively charged ions are known as cations (K^+, Mg^{++}, Ca^{++}, Fe^{+++}, Mn^{++}, Zn^{++} and Cu^{++}) and the negatively charged ones are known as anions (NO_3^-, $H_2PO_4^-$, SO_4^{--}, CI^-, $HB_44O_7^-$and $HMoO_4^-$). The ions by roots and the liberation of ions from soil particles are two independent processes whose common of ions from soil particles are two independent processes whose common poll is the soil solution.

Transfer of ions into roots involves the exchange of cations. Likewise, anions are absorbed by roots in exchange of OH-and HCO3- ions. Further roots exhibit selective action in absorbing several ions from soil solution which is referred to as the internal regulation of absorption or selective absorption. External factors governing absorption by roots are the concentration of the soil solution and the relative preponderance of individual ions in the root region.

Structure and composition of three mineralogical types of Clay; A- Kaloinite; B- Montmorillonite; C- Illite (From Donehue et al., 1971)

Fixation of Ions in Soil

It has been estimated that approximately 5% of the total nitrogen in surface soils and 60% of total nitrogen in sub-soils are held as non-exchangeable (fixed) ammonium. This property of soils to retain NH4+ ion is known as ammonium fixation, the mechanism of which is highly complicated. The ammonium ion replaces other cations in soil such as Ca, Mg, Na and H in the expanded lattice of clay minerals. During this process, the lattice of the clay minerals contracts and traps ammonium ions in a non-exchangeable (fixed) form. Potassium fixation also takes place by a similar mechanism. Most of the phosphorus added to soil is also fixed as insoluble compounds of iron, aluminium, orthocalcium, phosphates and apatite. This situation can be remedied if soils receiving fertilizer phosphorus are amended with organic matter. The decomposing organic residues form citrates, tartarates, acetates, oxalates, malates, malonates and other organic anions which release fixed phosphorus.

SOIL WATER

The water holding capacity of a soil is governed by the porosity or soil structure. The field capacity of a soil is 'the amount of water held in the soil after the excess gravitational water has drained away and after the rate of downward movement of the water has materially decreased.

This can be measured by saturating the soil with water followed by draining for 2-3 days under normal conditions which is then expressed as the percentage of water in the dry weight or volume of the soil.

The supply of water to plant roots depends not only on this field capacity of soil but also on water tension or suction tension (pF value) created by the suction of water to the surface of soil particles. The pF a which plants begin to wilt is called the 'wilting point'.

This wilting point is reversible but later the plant dies at the irreversible stage. Water is bound if pF values are higher than 4.0-4.2 and thus the water available to plants is the field capacity minus this value. Therefore, in clay soil water is unavailable to plants to appreciable extent even though saturated with it. Movement of water by capillarity which is often lower saturated with it. Movement of water by capillarity which is often lower in lay soils and higher in sandy soil influences the availability of water to plants.

Soil aeration is related to soil water content. The difference between the sum of pore space in a given sample of soil and the water content accounts for the air-filled space. Adequate aeration of soil can be achieved by a minimum air-filled pore space of 10 per cent by volume. Clay soil with 45 per cent of water has the lowest aeration among different soil types.

The occurrence of water in soil can be three types¯gravitational, capillary and hygroscopic. After heavy rain or irrigation. gravitational water is drawn through the soil. Upon drainage of excess water, water is retained at field capacity in soil pores and this water is known as capillary water. Hygroscopic soil water relates to that water absorbed by dry soil from an atmosphere of high relative humidity. Fungi are mostly tolerant to greater water stress than are bacteria. An example of microorganisms less tolerant to water stress is Nitrosomonas, followed by ammonifiers such as Clostridium and Penicillium.

SOIL ATMOSPHERE

Major gases in soil are N_2, O_2 and CO_2 content of soil air (0.3-1.0 per cent by volume) is higher than that of the atmospheric air (0.03 per cent by volume). This is due to the respiratory activity of soil microorganisms in which O_2 is consumed leading to lower O_2 content of soil atmosphere.

In submerged paddy soils, at first anaerobiosis reduces O_2 but very soon rice roots excrete O_2 into the rhizosphere. However, reducing conditions prevail, depending on the depth.

In other situations where water logging is a perennial feature, plants adapt their metabolism. For instance, instead of sugars, plants may end up accumulating malate. Under highly anaerobic conditions, toxic substances such as ethane, methane, hydrogen sulphide, cyanide, butyric acid and a number of other fatty acids may accumulate causing harmful effects to plant growth.

METHANE EMISSIONS FROM LOWLAND RICE CULTIVATION

Flooded rice fields are generally anoxic and thus provide excellent habitat for Methanogenic microorganisms. Methane (CH_4) absorbs long-wave radiation and therefore is a green house gas that leads to the warming of the earth's surface.

Together with CO_2 and nitrous oxide, methane traps the thermal radiation from the earth's surface and influences the atmospheric chemistry of ozone in the troposphere as well as the stratosphere through photochemical reactions. The present estimates of CH_4 emissions from wetland rice cultivation are placed around 25-17 Tg CH_4 per year. Recent studies however point out that India rice soils emit around 4 to 6 Tg methane per year, the upland rice cultivation registering lower values than lowland rice cultivation. Other estimates put the values around 20-150 Tg per year. It appears that the variability may be due to varietal differences of rice planted and to different agronomic practices followed besides the factor contributed by erratic rainfall.

One of the mitigation options found out by experiments is to grow rice cultivars that emit low levels of CH_4. A single midseason drainage may reduce seasonal methane emission rates by about 50 per cent but this practice may generate nitrous oxide, another green house gas.

Application of gypsum that releases sulfide inhibits CH_4 formation and similarly application of rock phosphate of single superphosphate does the same trick. Likewise, foliar application of technical grade Organochlorine insecticide hexachlorocyclohexane (HCH) to rice plants retards CH_4 production. Most of the options to reduce methane emissions in rice cultivation are however location specific and have to be adopted cautiously so as to avoid lowering rice production.

SOIL PH

The negative logarithm of the activity of H^+ in solution is denoted by pH which indicates the acidity/alkalinity of soil when measured in an aqueous or KCI extract. In the latter case, some of the H^+ ions absorbed to soil colloids are also replaced by K^+ and hence the pH indicated would be lower than water extract. The buffering capacity of soil, depending on the type of soil, also determines soil pH. Soils with high organics or inorganic colloids exert great buffering capacity because of saturation with cation species. The pH value of soils vary from 3-10. The acid sulphate and podzolic soils are low in pH whereas calcareous and alkali soils are very high in pH due to strong bases such as Na^+ or K^+ and weak acids such as $HCO3^-$.

The pH os soil determines the mineral content as well as microbial composition. High pH release K^+, Mg^{2+}, Ca^{2+}, Mn^{2+}, Cu^{2+}, and Al^{3+} by weathering processes of soil whereas low pH favours solubility of salts including carbonates, phosphates and sulphates. Generally, fungi are predominant in the rhizosphere under low pH conditions (< 5.5) and beneficial nitrogen fixing microorganisms are favoured by neutral pH. Plants adapt to soil pH and this depends upon species. Plants like rye, oats, potatoes and wheat adjust to pH range of 4.1-7.4. On the other hand, lupines have pH requirements in the limited range of 4.1 to 5.5. Liming soil is a well-known practice to correct pH of soils for plant growth. Generally, soils of temperate regions tend to become acidic mainly due to the production of H^+ ions organic matter decomposition.

Saline soils (Solonchak) contain an excess of neutral salts such as chlorides and sulphates of Na^+ and Mg^{2+}. When ground water rises followed by evaporation in semi-arid conditions, salts surface as white patches.

Alkali soils or solonetz (pH 7.5-10.0) are characterized by the presence of Na_2CO_3 and $NaHCO_3$. Sometimes alkali soils are black due to mixture of humic substances.

The total ion concentration of these soils are not conducive to plant growth. The degree of salinity is usually measured in the water extracts of soils in terms of electrical conductivity. This is expressed as mm hos/cm which is the reciprocal of electrical resistance. As the salt content increases the electrical conductivity also increases. Salt affected soils are common in many countries with conductivity values in the range of 2-20 mm hos/cm in saturated extracts.

5

Soil Fertility

Soil samples (from 0-20 cm depth) were taken from each experimental plot immediately after each harvest. Equal amounts (3 kg) were put into clay pots (about 5 litre capacity) for a biological test of overall soil fertility. Three seeds of maize were planted. After 5 weeks the maize plants were removed from the soil, washed to remove soil from the roots and allowed to dry for 1 hour. The total fresh biomass and the maize roots were weighed (Maria Elena Gomez 1993 personal communication).

Soil Analysis

Samples of soil (from 20 cm depth) were taken from each plot after the 12 month harvest. The 18 samples corresponding to the mulching treatments were bulked and 6 sub-samples taken for analysis. The samples from the 18 non-mulching treatments were treated in the same way. The sub-samples were analysed for pH, N, P, K and carbon by standard methods (AOAC 1985). Estimates were made of the populations of fungi, bacteria and actinomycetes according to "Standard methods of analysis for Soil, Plant tissue, Water and Fertilizer" (Philippine-Los Banos, Laguna 1980) with asparagine-manitolagar medium for bacteria, glycerol-agar for actinomycetes and peptone-dextrose-agar plus rose bengal and streptomycin for fungi.

SOIL COMPACTION AND BUILDING

Soil compaction is defined as the method of mechanically increasing the density of soil. In construction, this is a significant part of the building process. If performed improperly, settlement of the soil could occur and result in unnecessary maintenance costs or structure failure. Almost all types of building sites and construction projects utilize mechanical compaction techniques

Soil density is used almost exclusively by the transportation industry to specify, estimate, measure, and control soil compaction. Soil density can be easily determined via weight and volume measurements. The objective of compaction is to stabilize soils and improve their engineering properties. Source: US Dept. of Transportation

There are five principle reasons to compact soil:-

- Increases load-bearing capacity-
- Prevents soil settlement and frost damage-
- Provides stability-
- Reduces water seepage, swelling and contraction-
- Reduces settling of soil-source Soil Compaction Handbook.

Porosity

Bulk Density is an indirect measure of soil pore space. In fact, there is a formula: 100-((B.D. ÷ Particle Density) X 100)=% pore space or POROSITY. Under field conditions, pore spaces are occupied at all times by air and water. "Tortuous pathways" best describes soil pores. Porosity, when expressed as a percent, is the same thing as percent pore space. Soil particles have irregular shapes, and thus the spaces or pores between them vary irregularly in size, shape and direction. Sandy soils have large continuous pores, while clays have small pores which transmit water slowly. Clays, however, contain more pore space than sandy soils, because of the pores inside of the soil peds. To growing plants, pore sizes are of more importance than total pore space. We will be discussing this more in the unit on soil water.

Particle Density

In the laboratory you will determine porosity and particle density for the sandy soil used previously. Particle density will also be determined for this soil. Particle density only takes into account the volume occupied by the solid particles. It excludes the volume occupied by air and water. Since a large portion of most soils is composed of particles derived from quartz minerals, the particle density of most soils is near 2.6 g/cc, which is the density of quartz. Variations in the particle density are due to the presence of heavier minerals like iron oxides or lighter organic components.

Significance of Porosity

Above, a weakly developed soil is on the left and a well developed soil is on the right. Where clay accumulation is minimal, the macro pore space is about the same with increasing depth. However, where clay has moved into pores, the amount of macro pores and the rate of water movement both decrease. Soils with clay films will have slow water flow through the Bt horizon.

Some soils are difficult to aerate by plowing due to the turf cover. For lawns, playgrounds, or football fields, aeration is accomplished by making small holes in the turf with solid tines or using hollow tines to pull out a small core. Later the holes can filled with sand to promote air exchange. This is called *aerifying* and *topdressing*. Many home lawns are compacted because of the construction activities when the home was built or due to people walking or riding on the turf when the soil is wet. Aerification will improve the turf growth on these lawns.

Soil Temperature Regimes

The temperature of a given soil at a given time is dependent upon the gains and losses of heat energy. Generally, dark surfaces will absorb more heat than light surfaces. However, the amount of water in the soil is an important factor. Losses of the absorbed heat are by radiation back into the atmosphere as long-wave radiation, heating the air and cooling the soil.

These soils warm quickly in the spring due to their dark surface. Soil temperature regimes are used to classify soils. They are defined according to the average annual soil temperature in the root zone. The use of soils for agriculture and forestry is closely related to soil temperature, due to the specific requirements of plants. Over most of the earth, daily soil temperatures below 50 cm deep seldom change. To approximate the mean annual soil temperature, 2 º F is added to the mean annual air temperature. If we do this for the annual temperature map of Minnesota and follow the 45ºF line, which would be 47º soil temperature, the state is divided along a line from the Twin Cities to the South Dakota border, and soils are frigid to the north and mesic to the south.

INTERPRETATION OF ELECTRICAL CONDUCTIVITY OF SOILS

In actuality, the interpretation of EC of a soil or media must be made considering the plant(s) to be grown. The EC of the soil has little direct detrimental effect on sandy mineral soils or on media. However, EC directly affects plants growing in the soil or media. The impact of EC on plants is also directly affected by water management.

Salt Index Use and Calculation

As EC increases, more attention to water management is needed to prevent salinity from adversely affecting plants. The Extension Soil Testing Laboratory uses a 2:1 solution:soil ratio with which to determine EC. Many states use a saturated paste extract. This saturated paste method is more time consuming than the 2:1 extraction, and results in inadequate amounts of solution in Florida's sandy soils. The conversion from the 2:1 extraction result to the saturated paste result, termed salt index, is easy and accurate.

EC (salt index) = EC (2:1) x 8.

In general, when the soil EC (2:1, water:soil) exceeds 0.25 dS/m (or 0.25 x 8 = 2.0 dS/m salt index), many plants experience stress due to salts. Other plants (e.g., bermudagrass) are quite tolerant to salts. Due to this species-dependent effect of EC, a listing of the effects of increasing EC on selected plants has been compiled (Hanlon et al., 1993).

EC Probe and Meter

There are a wide variety of EC probes and meters marketed. Those instruments using probes that are placed in the soil or media *in situ* (directly in the soil without taking a sample) are not considered in the following discussion. Since EC is a measurement of the conductivity of the soil solution, the measurement should be made under a controlled mixture of solution to soil (2:1). These conditions usually do not exist when direct reading instruments are used.

Electrical Conductivity Probe

The probe consists of a tube, usually of plastic, into which electrodes have been installed. Two common electrode arrangements are: 1) two plates; or 2) a rod located concentrically in a ring. In either case, the electrodes are held a specific distance apart. The gap between the electrodes is filled with the water sample or filtered solution from a soil sample, either by filling a reservoir or by placing the probe in the solution. When the probe is immersed in the solution, ions contained in the solution will permit electrical flow from one electrode to another. If a large number of ions are present (salty conditions), then the EC of the sample will be higher than a sample with low number of ions (low salts).

Some older instruments require that a mixture of soil and solution be packed into the electrode receptacle. While readings from these older instruments may be adequate, accurate results are harder to obtain because readings are affected by the sample packing method.

Inspection of the probe should include insuring that the probe is clean and free of debris, and that all electrical connections are in place. Some probes may experience a buildup of corrosion on the electrodes with time. Indication of corrosion, if not directly visible, is indicated by the constant need to reset the instrument because of drifting, often in one direction as corrosion progresses.

Electrical Conductivity Meter

Meters for EC are extremely reliable. Meters from the 1950s which use a "cat's eye" tube are still in use. Newer meters using digital displays are often susceptible to failure of one or more segments of the digital display, usually related to corrosion within the instrument.

As with failures associated with pH meters, EC meters will always provide an EC, unless the problem results in total instrument failure. It falls upon the operator to insure that the reported EC is an accurate measurement of the conductivity of the sample.

LABORATORY PROCEDURES

The following procedures are used at the ESTL and are based on good laboratory procedures with sufficient quality control measures to ensure that pH and EC readings are accurate and reliable. While equipment may vary among County Extension Laboratories, procedures should be developed that directly parallel those used by the ESTL.

Soil Scooping Technique

The ESTL uses a scoop (that is, a volume\ measurement) for both pH and EC determinations. The scoop is a plastic and metal device which may be obtained from a commercial manufacturer (e.g., Custom Laboratories, Orange City, FL). Alternately, the scoop may be constructed from locally available materials, such as measuring spoons or coffee scoops. The intent is to use a consistent volume of soil and water.

The soil scooping technique requires practice, despite its unsophisticated appearance. The technique depends upon uniform actions by the technician from sample to sample to produce consistent packing of soil into the scoop. This consistency can be directly measured by repeatedly scooping the same soil and weighing each scoop. Weights should be uniform within each scoop (volume) and soil-sample combination. Weights will vary from soil to soil, especially when there is a noticeable difference in soil texture.

Sample Handling and Preparation

The sample should be air-dried and passed through a 2-mm sieve before scooping.

Procedure

1. Dip the scoop into the center of the soil sample and fill the scoop with a twisting motion so that extra soil is mounded above the rim of the

scoop. Do not press the scoop or force the soil against the side of the container (Jones, 1980).

2. Strike the handle near the scoop three times with a plastic rod to settle soil particles.
3. Level the scoop with the plastic rod. Strike off all excess soil above the rim of the scoop in a single stroke so that the soil is not compacted into the scoop.

Estimation of Soil Texture

Knowledge of soil texture is useful in the recommendation of lime. Local soil conditions may be quite uniform so that little differences may be found throughout the county. However, such uniformity is expected to be rare in most counties. After the soil has been air-dried and sieved, a small amount of dry soil should be moistened and rubbed between the forefinger and thumb. An estimation of the texture is made by comparing the"feel" of the sample to that of a set of soil samples of known texture.

Soil pH (2:1 V/V)

This procedure uses a 20-cc (~25-g) soil scoop and 40 mL of pure water to obtain a 2:1 water-to-soil ratio. Most problems with this procedure are associated with the glass or calomel electrodes.

However, sample pH may also be affected by contaminated water, by microbial activity if samples are allowed to sit for several hours before determining pH, or by improper scooping techniques.

1. *Standard Solutions:* Obtain commercial standard solutions of pH 4.00, 7.00, and 10.00.

   Sample Handling and Preparation: The soil sample should be air-dried and passed through a 2-mm sieve. Irrigation water samples require no preparation.

   Procedure: 1. One scoop of soil to a 3-oz plastic cup using a 20-cc (~25-g) scoop.
2. Add 40 mL of pure water to each cup using an automatic pipette or suitable volumetric container. Stir with a glass rod and let the sample stand for 30 min.
3. Standardize the pH meter.
4. Stir the sample again immediately before measuring the soil pH. Do not place the electrode(s) directly in the sand layer at the bottom of the container. The electrodes should be positioned in the solution just above the sand layer.

5. Record pH to the nearest 0.1 pH unit (suggested format of XX.X).

Electrical Conductivity

This test, often called "Soluble Salts," requires that 20 cc (~25 g) of soil be mixed with 40 mL of pure water, resulting in a water:soil ratio of 2:1. The 4-hr equilibration period provides time for some slowly-soluble constituents to approach solution equilibrium. Little error results from much longer equilibration times, but shorter time periods might introduce inconsistent results for some samples.

Standards

A solution of 0.005 N KCl has an electrical conductivity of 720 ± 1 dS/m (mmho/cm) at 25°C. Use a commercially prepared solution as the reference solution.

Sample Handling and Preparation

- The soil sample should be air-dried and passed through a 2-mm sieve. Irrigation water samples require no preparation.

Procedure

1. Place 20 cc (~25 g) of soil in a plastic 3-oz cup.
2. Add 40 mL of pure water, stir, and allow to stand for 4 h.
3. Without stirring the sample, filter the solution through a Whatman No. 41 (11 cm) paper and collect the extract in a funnel tube or other suitable container. The intent is to remove the soil and other debris from the solution.
4. Standardize the conductivity meter.
5. Move the probe up and down in the solution several times to dislodge any bubbles on the electrode surfaces. Measure the electrical conductivity of the extract contained in the funnel tube.
6. Rinse the interior and exterior of the probe with pure water between samples. Remove any excess water from the exterior of the probe by blotting with a tissue.
7. Record all meter readings as displayed. Note that some meters use a floating-point display while others use a reading which depends upon a switch setting.

pH Meter

Instructions for the proper use of a pH meter, including electrode care and instrument-specific settings, are given in this subsection. While instrument settings may vary among meters, daily operation of pH meters is relatively standard. Most problems with pH instruments originate within the electrode(s) or in the electrical connections to the instrument. Prevention of early electrode deterioration is best accomplished by following the information in sections B and C that follow. Use of only one standard buffer solution is not adequate for proper calibration. Do not reuse a buffer solution which has been left out overnight or has been used to "store" electrode(s). It is advisable to maintain a supply of a reference soil sample to be used as a check sample with every set of samples. Early detection of problems is the best method of avoiding inaccurate results.

Calibration of Electrode(s)

1. Rinse electrode(s) with pure water and blot dry with a tissue or clean paper towel.
2. Using commercially prepared buffer solutions of pH 4.00, 7.00, and 10.0, pour approximately 30 mL into labeled, 3-oz. plastic cups. Do not reuse buffer solutions in which electrode(s) have been immersed for daily storage.
3. Set the meter to operate in the "pH" mode.
4. Place the electrode(s) into new buffer solution (pH 7.00) and allow the electrode(s) to equilibrate. A stable reading should be obtained within about 30 seconds. Readings which are not stable after 1 minute indicate that the electrode(s) may be malfunctioning.
5. Using the Calibration" (often labelled "Standard") knob, adjust the pH-meter display to read 7.00. Do not move (or press) either the "Temperaturer" or "Slope" adjustments at this time.
6. Place the meter in the "Standby" mode. Some meters do not have a standby mode.
7. Rinse the electrode(s) into a waste cup with pure water and blot dry as above.
8. Place the electrode(s) in new buffer solution (pH 4.00) and allow the electrode(s) to equilibrate with the buffer solution.
9. Set the meter to operate in the "pH" mode.
10. Using the "Temperature" knob (sometimes labelled "Slope"), adjust the pH-meter display to read 4.00. Do not readjust the "Calibration" setting at this time.
11. Repeat Steps 6 through 9 using the pH 10.0 buffer solution. No settings should be changed as the pH 10.0 buffer is being read.

12. Repeat Steps 3 through 11 until readings of 7.00, 4.00, and 10.0 are obtained without adjusting the instrument. These standards should be set ±0.05.
13. Using a reference soil sample, read the pH and determine if the pH reading agrees with the "known" reading of the reference soil. Use of a reference soil sample, a soil that can be analyzed with every sample set, is strongly recommended. Use of a reference soil is an excellent quality assurance measure and verifies that the electrode(s)/meter are functioning correctly in a soil solution and not just in standard buffers. The reference soil sample should be read once about 20 samples. If the reading drifts by more than ±0.2, then restandardize the meter.
14. Standardize the instrument according to the above procedure after every 50 samples. If the instrument has shown drift of 0.05 pH units or more, reread the last few samples to verify the accuracy of the recorded readings.

Daily Electrode Storage

1. Place the meter in the "Standby" mode.
2. Wash the electrode(s) with pure water.
3. Immerse the electrode(s) in pH 7.00 buffer solution.

Electrode Servicing

1. If the pH reading drifts, replace the glass electrode. Wait until the reading from the new electrode is stable. Usually, new electrode(s) should be kept in pH 7.00 buffer solution overnight before use.
2. After use, wash the electrode(s) with pure water.
3. Immerse the electrode(s) in pH 7.00 buffer solution.

Electrical Conductivity Meter

Conductivity measurements must be made on solution samples only. The most common problem with conductivity meters concerns failure of the electrodes (mounted within the hollow plastic probe) to make proper contact with the solution.

The probe must be kept clean by adequately flushing the interior of the probe with pure water. When analyzing a solution, insure that the solution covers the bottom 3 to 5 cm (1 to 1.5 inches) of the probe. Agitation of the probe in the unknown and standard solutions is required to obtain reproducible results. Agitation insures complete wetting of the internal electrodes by removing any air bubbles using the solution of interest.

Calibration

1. Turn ON the "Supply" or "Power" switch.
2. Turn OFF the "Temperature Correction" switch (back panel).
3. Set the "Scale" wafer switch, if any, to read mmho/cm.
4. Place the conductivity probe into a standard solution of 0.005 M KCl. Agitate the probe using an up-and-down movement to create better solution to probe contact.
5. After agitation, the meter should read 0.72 ± 0.04 mmho/cm (0.72 ± 0.04 dS/m).
6. Agitate the probe again and reread the standard solution. Both the first and second readings should be the same value if the probe is in good contact with the solution.
7. Wash the probe with pure water on both inner (inject wash water through the hole at the top of the probe) and outer surfaces. Blot the probe dry. Do NOT rub the outer surface of the probe with the tissue paper.
8. Read the reference soil sample and verify that the current reading agrees with the known value for the reference soil.
9. Record the entire meter reading including location of the decimal point.
10. Return the "Scale" wafer switch to the original position before reading the next sample.

SOIL ELECTRICAL CONDUCTIVITY VARIABILITY

Soil electrical conductivity (EC) is a property of soil that is determined by standardized measures of soil conductance (resistance^{-1}) by the distance and cross sectional area through which a current travels. Traditionally, soil paste EC has been used to assess soil salinity (Rhoades et al., 1989), but now commercial devices are available to rapidly and economically measure and map bulk soil EC across agricultural fields. The Veris® 3100 (Veris Technologies, Salina, Kansas) measures EC with a system of coulters that are in direct contact with the soil. The EM38 (Geonics, Limited, Mississauga, Ontario, Canada) induces a current into the soil with one coil and determines conductivity by measuring the resulting secondary current with another coil. Both sensors have been demonstrated to give similar results (Suddeth et al., 1999)

The movement of electrons through bulk soil is complex. Electrons may travel through soil water in macropores, along the surfaces of soil minerals (i.e. exchangeable ions), and through alternating layers of particles and solution (Rhoades et al., 1989). Therefore, multiple factors contribute to soil EC variability, including factors that affect the amount and connectivity of soil water (e.g. bulk density, structure, water potential, precipitation, timing of measurement), soil aggregation (e.g. cementing agents such as clay and organic matter, soil structure), electrolytes in soil water (e.g. salinity, exchangeable ions, soil water content, soil temperature), and the conductivity of the mineral phase (e.g. types and quantity of minerals, degree of isomorphic substitution, exchangeable ions). Despite the multiple causes of EC variability, bulk soil EC measurements have been related to individual factors that limit soil use and productivity such as salinity (De Jong et al., 1979; Rhoades and Corwin, 1981), clay content at a depth of 15-m in New Wales, Australia (r^2 = 0.78; Williams and Hoey, 1987), depth of sand deposition along the Missouri River (r^2 = 0.73-0.94; Kitchen et al, 1996), depth to claypan in Missouri (r^2 = 0.73; Doolittle et al., 1994), and soil moisture content (r^2 = 0.96; Kachanoski et al., 1988).

If soil EC maps have utility in production agriculture, 1) EC must be spatially structured, 2) spatial patterns must have temporal stability, and 3) EC must be related to factors of agronomic importance. The objective of this research was to determine the nature and causes of EC variability for several fields in Kentucky. Geostatistical analyses were conducted to examine the spatial and temporal variability of EC variability. Transect studies were conducted to determine the causes of EC variability.

SOIL AND ITS FERTILITY

An experiment was carried out with sugar cane at the Goat and Rabbit Research Centre, Hatay, Vietnam, from January 1993 to December 1995 to study the effect of row width (75, 100 or 150 cm), plant material (stem cuttings or tops), and removing or leaving the dead leaves, on biomass yield and juice quality (°Brix). Yield of cane stalks was higher when stem cuttings, rather than tops, were used as planting material.

The effect was most pronounced in the first harvest year and became less with succeeding harvests. The cost of seed was also reduced by 9-10% compared with the traditional way used by the farmers in North Vietnam. Increasing population density by reducing inter-row spacing to 75 cm led to higher yields of biomass with no reduction in Brix or extraction rate of the juice.

Mulching with the dead leaves increased yields, the difference being more marked in the second than in the first year and even more apparent in the third year (6.3, 20 and 30 % increases for mulching in plant, 1st and 2nd ratoons, respectively). Mulching also improved soil fertility and increased the amount of carbon sequestered in the soil, indicating that mulching is a positive alternative to the traditional way of removing the dead leaves. Soil fertility increased steadily with increasing ratoons indicating that the growing of sugar cane does not exploit soil nutrients but, in contrast, has beneficial effects on growth of subsequent crops.

In recent years a considerable amount of research has been directed at increasing biomass quantity and quality of agricultural crops used as animal feed. Sugar cane is a perennial crop which can be used for animal feed as well as sugar production (Preston and Murguetio 1992). It has a high leaf area index and a high photosynthetic efficiency under strong sunshine, more than any other crop in the tropics (Bassham 1978).

The individual and combined effects of certain management practices—planting date, row spacing, planting depth, fertilizer rate, pest control and irrigation — have a great impact on the growth and yield of sugar cane. The number of plants at harvest time is a major determinant of biomass yield, and a density of 50-70,000 plants/ha was recommended by Tran Van Soi (1988). This was a general recommendation to account for factors such as variety, branching ability, climate and soil conditions.

If the plants are too close, there may be too many shoots which will reduce the efficiency of the parent plants and class 1 branches while too large a space between rows will lead to a waste of the area and of solar energy (Duong Duc Thang 1991).

In the hilly land of Vietnam, higher production was obtained with the same plant density but arranged in narrower rows (Nguyen Huy Uoc 1987) compared to the traditional method of planting. Other researchers also reported higher yields from narrow spacing (90 and 60 cm) compared with wide spacing (130, 140 and 180 cm) (Sharma 1982; Irvine *et al* 1984; Singh and Singh 1984; Gonzalaz *et al* 1989; Arvind Misra *et al* 1990).

Soil organic matter is an important feature of soil fertility. Ways of increasing soil organic matter are by growing a cover crop (green manure) or by mulching with compost or crop residues. These practices also help to control weed growth. Mulching has been shown to increase the yield of tomatoes and tuber plants (Kaniszewski 1994; Pakyurek *et al* 1994, Khalak and Kumaraswamy 1993; Shin *et al* 1993). The method gave higher yields in fruit trees, vegetables and crops such as mango, sweet pepper, bananas, guava and maize (Farre *et al* 1993; Siwek *et al* 1994; Singh and Singh 1992; Sarad-Gurung *et al* 1994; Sandhu *et al* 1992).

Mulching by crop residues was superior to polyethylene sheets (used to control weeds) in terms of incremental cost-benefit ratios (Khalak and

Kumaraswamy 1993). There are several reports of improvement of soil fertility as a result of mulching with crop residues (Sandhu *et al* 1992; Kitou and Yoshida 1994; Arzeno 1992).

The hypotheses to be tested in the present study were:

- Reducing row distance would lead to increased yield of biomass
- Planting with stem cuttings rather than the growing point would increase the longevity and the yield of sugar cane
- Returning the dead leaves to the soil (mulching) would increase biomass yield and soil fertility

MATERIALS AND METHODS

Location: The experiment was carried out from January 1993 to December 1995 at the Goat and Rabbit Research Centre in Bavi district in a hilly area some 60 km North-West of Hanoi with an average slope of 3-8°.

The soil composition at the experimental site was as follows: pH 5.3, K_2O 0.06%, P_2O_5 0.09%, N 0.14%. In general, the soil is of low fertility and the organic matter has been reduced to very low levels by erosion. Drought occurs frequently in the area.

Treatments and Design: The treatments in a split-plot design with three replications were:

- Planting distances of 75, 100 or 150 cm between rows
- Planting material: Tops (growing points of sugar cane) or stem cuttings (the stalk was cut into pieces of 50 cm)
- Mulching (returning the dead leaves to the soil surface) or no mulching (taking away the dead leaves)
- Harvesting at 10 months or 12 months after planting

Row spacing was the main plot and plant material and mulching were the sub-plots. The area for each experimental plot was 90, 80 and 75 m² for row spacings of 150, 100 and 75 cm, respectively.

The size of the recorded area was 60 m² for all treatments. The total number of plots was 36, giving a total experimental area of 3,569 m² (2,940 m² for recording data). The plant material was set in double continuous rows with 8, 10 and 12 tonnes/ha for 150, 100 and 75 cm spacing between rows, respectively.

Cultural Practices: Fertilizer was applied according to the normal practice by farmers in the area. The amounts of cattle manure and chemical fertilizers applied.

LEMON

Citrus Limon

The leading acid citrus fruit, because of its very appealing colour, odour and flavour, the lemon, *Citrus limon* Burm. f. (syns. C. *limonium* Risso, C. *limonia* Osbeck, *C. medica* var. *limonium* Brandis), is known in Italy as *limone;* in most Spanish-speaking areas as *limon, limon agria, limon real,* or *limon frances;* in German as *limonen;* in French as *citronnier;* in Dutch as *citroen.* In Haiti, it is *limon France;* in Puerto Rico, *limon amarillo.* In the Netherlands Antilles, *lamoentsji,* or *lamunchi,* are locally applied to the lime, not to the lemon as strangers suppose. The lemon is not grown there. Several lemon-like fruits are domestically or commercially regarded as lemons wherever they are grown and, accordingly, must be discussed under this heading. These include: Rough lemon (C. *jambhiri* Lush.), Sweet lemon *(C. limetta* Risso), 'Meyer' (lemon X mandarin hybrid); 'Perrine' (lime X lemon hybrid); 'Ponderosa' (presumed lemon X citron hybrid), qq.v. under "Varieties".

The true lemon tree reaches 10 to 20 ft. (3-6 m) in height and usually has sharp thorns on the twigs. The alternate leaves, reddish when young, become dark-green above, light-green below; are oblong, elliptic or long-ovate, 2 1/ 2 to 4 1/2 in (6.25-11.25 cm) long, finely toothed, with slender wings on the petioles. The mildly fragrant flowers may be solitary or there may be 2 or more clustered in the leaf axils. Buds are reddish; the opened flowers have 4 or 5 petals 3/4 in (2 cm) long, white on the upper surface (inside), purplish beneath (outside), and 20-40 more or less united stamens with yellow anthers. The fruit is oval with a nipple-like protuberance at the apex; 2 3/4 to 4 3/4 in (7-12 cm) long; the peel is usually light-yellow though some lemons are variegated with longitudinal stripes of green and yellow or white; it is aromatic, dotted with oil glands; 1/4 to 3/8 in (6-10 mm) thick; pulp is pale-yellow, in 8 to 10 segments, juicy, acid. Some fruits are seedless, most have a few seeds, elliptic or ovate, pointed, smooth, 3/8 in (9.5 mm) long, white inside.

Origin and Distribution: The true home of the lemon is unknown, though some have linked it to northwestern India. It is supposed to have been introduced into southern Italy in 200 A.D. and to have been cultivated in Iraq and Egypt by 700 A.D. It reached Sicily before 1000 and China between 760 and 1297 A.D. Arabs distributed it widely in the Mediterranean region between 1000 and 1150 A.D. It was prized for its medicinal virtues in the palace of the Sultan of Egypt and Syria in the period 1174-1193 A.D. Christopher Columbus carried lemon seeds to Hispaniola in 1493. The Spaniards may have included lemons among the fruits they introduced to St. Augustine. They were grown in California in the years 1751-1768. Lemons were reported to be increasingly planted in northeastern Florida in 1839. Because of heavy imports from Sicily, commercial culture in Florida and California was begun soon after 1870 and grew to the point where 140,000 boxes were being shipped out of Florida alone.

The small Florida industry was set back by a freeze in 1886, the susceptibility of the lemon to scab, and the unfavorable climate for curing the fruit, and also competition from California. Following the devastating freeze of 1894-95, commercial lemon culture was abandoned in Florida. Not until 1953 was interest in lemon-growing revived in Central Florida to take advantage of the demand for frozen concentrate and for natural cold-press lemon oil. At that time, Florida was importing lemons from Italy for processing. Plantings grew to 8,700 acres by 1975. Freezes caused 50% reduction by 1980. Still, in 1984, Florida exported $2 million worth of lemons.

In the meantime, Arizona had developed lemon orchards, though on a smaller scale than California. In the 1956-57 season, California produced 11 million gallons (42 million liters) of frozen lemon concentrate while Florida's output was still very small.

California and Arizona became the leading sources of lemons in the western hemisphere. In recent years, California has produced nearly double the crop that can be profitably marketed fresh or processed. Foreign competition has increased and many California growers have destroyed their lemon groves or topworked the trees to oranges, but new cultural techniques making summer production possible may reverse the trend.

Guatemala has in the past 2 decades developed commercial lemon culture, primarily to produce the peel oil for its essential oil industry and secondarily for the purpose of dehydrating the fruit and preparing a powder for reconstituting into juice. Southern Mexico, too, is now a major grower of lemons, also primarily for lemon peel oil. Lemons are rarely grown for the fresh fruit market in Latin America. In South America, Argentina leads in lemon culture with Chile a distant second. Among the world's leading lemon growers and exporters are Italy, Spain, Greece, Turkey, Cyprus, Lebanon, South Africa and Australia. Lemons can be grown only at medium and high elevations in the Philippines.

Varieties: With the resumption of lemon-growing in Florida, workers at the Citrus Experiment Station, Lake Alfred, began a search for the most suitable cultivars, whether in dooryards, or in the United States Department of Agriculture planting at Orlovista, or the Lake Alfred collection. By late 1950, 200 selections had been brought together from various parts of the United States. Of these, 40 were budded onto 30-year-old grapefruit trees on rough lemon rootstock on the Minute Maid property at Avon Park. Two selections grown elsewhere were included in the studies-evaluation for thorniness, cold-and disease-susceptibility, sizes, juiciness, flavour, number of segments and seeds, yields, and quality of peel oil. The majority of the selections were judged undesirable; only a few showed promise for processing and fresh fruit marketing purposes. For processing, 'Villafranca' rated highest, followed by 'Eustis', 'Bearss', 'Perkin' and 'Avon'. Any of these, properly harvested and cured would be suitable for marketing fresh. Libby, McNeil & Libby, when planning for their lemon orchard at Babson Park, Florida, about 1948, tested varieties from all major lemon-producing areas of the world and chose 'Bearss' as rating highest in quality and quantity of juice, which was their chief concern at the time. In 1960, they added marketing of the fresh fruit and found the 'Bearss' equally desirable for this purpose.

The following are brief descriptions of most of the better known cultivars of true lemons and of lemon-like fruits that are accepted as lemons in home or commercial usage, and a few of the lesser-known.

'Armstrong' ('Armstrong Seedless')–a sport discovered in a private grove at Riverside, California, about 1909. Patented in 1936 by Armstrong Nurseries. Resembles 'Eureka' except that it usually bears seedless or near-seedless fruits. If planted among other lemon trees will occasionally have a few seeds.

'Avon'–first noticed as a budded tree in Arcadia, Florida. A budded tree propagated from the original specimen around 1934 was planted in the Alpine Grove in Avon Park; it produced heavy crops of fruits highly suitable for frozen concentrate. It, therefore, became the source of budwood for commercial propagation by Ward's Nursery beginning in 1940.

'Bearss' ('Sicily', but not the original introduction by Gen. Sanford in 1875, which has disappeared)–a seedling believed to have been planted in 1892, discovered in the Bearss grove near Lutz, Florida, about 1952. Closely resembles 'Lisbon'. It is highly susceptible to scab and greasy spot and oil spotting. The tree is vigorous and tends to produce too many water sprouts. Nevertheless, it has been propagated commercially by Libby, McNeill & Libby since 1953 because the peel is rich in oil. It constitutes 20% of Brazil's lemon/lime crop.,

'Berna' ('Bernia', 'Vema', 'Vernia')–oval to broad-elliptic, with pronounced nipple, short neck; peel somewhat rough, medium-thick, becoming thinner in summer, tightly clinging. Seeds generally few or absent. Ripens mostly in winter; fruits keep well on tree until summer but become too large. Tree is vigorous, large, prolific. This is the leading cultivar of Spain and important in Algeria and Morocco. It is too much like the 'Lisbon' to be of value in

California. In Florida, it has been found deficient in acid, low in juice, and too subject to scab.

'Eureka'–originated from seed taken from an Italian lemon (probably the 'Lunario') and planted in Los Angeles in 1858; selected in 1877 and budwood propagated by Thomas Garey who named it 'Garey's Eureka'. The fruit is elliptic to oblong or rarely obovate, with moderately protruding nipple at apex, a low collar at the base; peel yellow, longitudinally ridged, slightly rough because of sunken oil glands, medium-thick, tightly clinging; pulp greenish-yellow, in about 10 segments, fine-grained, tender, juicy, very acid. Fruits often borne in large terminal clusters unprotected by the foliage. Bears all year but mostly late winter, spring and early summer when the demand for lemons is high. Tree of medium size, almost thornless, early-bearing, prolific; not especially vigorous, cold-sensitive, not insect-resistant; relatively short-lived. Not suitable for Florida. Grown commercially in Israel. One of the 2 leading cultivars of California, though now being superseded by clonal selections with more vigor, *e.g.*, 'Allen', 'Cascade', 'Cook', and 'Ross'. 'Lambert Eureka' is a chance seedling found in 1940 on the property of Horace Lambert in New South Wales. It is vigorous and productive.

'Femminello Ovale'–one of the oldest Italian varieties; short-elliptic with low, blunt nipple; slightly necked or rounded at base; of medium size; peel yellow, finely pitted, medium-smooth, medium-thick, tightly clinging; pulp in about 10 segments, tender, juicy, very acid, of excellent quality, with few, mostly undeveloped, seeds. Fruits all year but mainly in late winter and spring; ships and stores well. The tree is almost thornless, medium-to very-vigorous, but highly susceptible to *mal secco* disease. This is the leading cultivar in Italy, accounting for 3/4 of the total lemon production, and 1/5 of the crop is processed as single-strength juice.

'Genoa'–introduced into California from Genoa, Italy, in 1875. Almost identical to 'Eureka'; ovoid or ovate-oblong with blunt nipple at apex; base rounded or slightly narrowed; of medium size; peel yellow, medium-thick, tightly clinging; pulp in 10-12 segments, melting, medium-juicy, with 29 to 51 seeds which are light-brown within. Tree is shrubby, nearly trunk-less, spreading, very thorny, cold-hardy. Grown commercially in India, Chile and Argentina.

'Harvey'–of unknown parentage; was found by Harvey Smith on the property of George James in Clearwater, Florida. Fruit much like 'Eureka'. Tree highly cold-tolerant, compatible with several rootstocks. Commercially propagated by Glen St. Mary Nurseries Company, near Jacksonville, Florida, since 1943.

'Interdonato' ('Special')–a lemon X citron hybrid that originated on property of a Colonel Interdonato, Sicily, around 1875; oblong, cylindrical, with conical, pointed nipple at apex, short neck or collar at base; large; peel yellow, smooth, glossy, thin, tightly clinging; pulp greenish-yellow, in 8 or 9 segments, crisp, juicy, very acid, faintly bitter. Very few seeds. Earliest in season; mostly fall and early winter. Tree vigorous, usually thornless, medium-resistant to *mal secco*; of medium yield; accounts for 5% of Italy's crop.

'Lisbon' (perhaps the same as 'Portugal' in Morocco and Algeria)–originated in Portugal, possibly as a selection of 'Gallego'; reached Australia in 1824; first catalogued in Massachusetts in 1843; introduced into California about 1849 and catalogued there in 1853; introduced into California from Australia in 1874 and again in 1875. Fruit almost identical to 'Eureka'; elliptical to oblong, prominently nippled at apex, base faintly necked; peel yellow, barely rough, faintly pitted, sometimes slightly ribbed, medium-thick, tightly clinging; pulp pale greenish-yellow, in about 10 segments, fine-grained, tender, juicy, very acid, with few or no seeds. Main crop in February, second crop in May.

Fruit is borne inside the canopy, sheltered from extremes of heat and cold. Tree large, vigorous, thorny, prolific, resistant to cold, heat, wind. Not well adapted to Florida. It is low-yielding and short-lived in India. Surpasses 'Eureka' in California. Has given rise to a number of clonal selections, particularly 'Frost', originated by H. B. Frost at the Citrus Research Station, Riverside, California in 1917 and released about 1950; also 'Prior Lisbon' and the more vigorous 'Monroe Lisbon'.

'Meyer'–a hybrid, possibly lemon X mandarin orange; introduced into the United States as S.P.I. #23028, by the agricultural explorer, Frank N. Meyer, who found it growing as an ornamental pot-plant near Peking, China, in 1908; obovate, elliptical or oblong, round at the base, occasionally faintly necked and furrowed or lobed; apex rounded or with short nipple; of medium size, 2 1/4 to 3 in (5.7-7.5 cm) wide and 2 1/2 to 3 1/2 in (6.25-9 cm) high; peel light-orange with numerous small oil glands, 1/8 to 1/4 in (3-6 mm) thick; pulp pale orange-yellow, usually in 10 segments with tender walls, melting, juicy, moderately acid with medium lemon flavour; seeds small, 8 to 12. Tends to be everbearing but fruits mostly from December to April. Tree small, with few thorns, prolific, cold-resistant; produces few water sprouts, and is only moderately subject to greasy spot and oil spotting. It is easily and commonly grown from cuttings.

Does well on sweet orange and rough lemon rootstocks; is not grafted onto sour orange because it is a carrier of a virulent strain of tristeza. Grown for home use in California; in Florida, both for home use and to some extent commercially for concentrate though the product must be enhanced by the addition of peel oil from true lemons, since that from 'Meyer' peel is deficient in flavoring properties. Has been fairly extensively planted in Texas and in Queensland, Australia, and New Zealand.

'Monachello' (Moscatello')–suspected of being a lemon X citron hybrid; elliptical, with small nipple and no neck, merely tapered at apex and base; medium-small; peel yellow, smooth except for large, sunken oil glands, thin, clinging very tightly; pulp in 10 segments, tender, not very juicy, not sharply acid. Bears all year but mainly winter and spring. Tree not vigorous, slow-growing, almost thornless, with abundant, large leaves; bears medium-well, resistant to *mal secco,* and has been extensively planted in Italy in areas where the disease is common.

'Nepali Oblong' (Assam', 'Pat Nebu')–originated in Assam; fruit resembles citron in some aspects; long-elliptic to oblong-obovate, with wide, short nipple; medium-large; peel greenish-yellow, smooth, glossy, medium-thick; pulp greenish-yellow in 11 segments, fine-grained, very juicy, of medium acidity, with few or no seeds. Everbearing. Tree large, vigorous, spreading, medium-thorny, prolific; foliage resembles that of the citron. Commercial in India.

'Nepali Round'–of Indian origin; round, without distinct nipple; juicy; seedless. Tree large, vigorous, compact, nearly thornless, medium-prolific. Successfully cultivated in South India.

'Perrine'–a Mexican lime X 'Genoa' lemon hybrid created by Dr. Walter Swingle and colleagues in 1909, but still a fairly typical lemon; it is lemon-shaped, with small nipple at apex, necked at base; of medium size; peel pale lemon-yellow, smooth, slightly ridged, thin, tough; pulp pale greenish-yellow, in 10 to 12 segments having thin walls; tender, very juicy, with slightly lime-like flavour but acidity more like lemon; seeds usually 4 to 6, occasionally as many as 12, long-pointed. Everbearing. Tree cold-sensitive but less so than the lime; resistant to wither tip and scab but prone to gummosis and other bark diseases. In the early 1930's, was extensively planted in southern Florida on rough lemon rootstock, but no longer grown.

'Ponderosa' ('Wonder'; 'American Wonder')–a chance seedling, possibly of lemon/citron parentage, grown by George Bowman, Hagerstown, Maryland around 1886 or 1887; appeared in nursery catalogs in 1900 and 1902; obovate, lumpy and faintly ribbed, slightly necked at base; large, 3 1/2 to 4 1/8 in (9-11 cm) wide, 3 1/2 to 4 3/4 in (9-12 cm) high; peel light orange-yellow, with medium-large oil glands, flush or slightly depressed; 3/8 to 1/2 in (1-1.25 cm) thick; pulp pale-green, in 10 to 13 segments with thick walls; juicy, acid; seeds of medium size, 30 to 40 or more, brown within. Everbearing. Tree small, moderately thorny; buds and flowers white or barely tinged with red-purple. More sensitive to cold than true lemons. Grown for home use and as a curiosity in California and Florida and in small-scale commercial plantings since 1948. Rather widely cultivated as an indoor potted plant in temperate regions.

'Rosenberger'–a clone found in a grove of 'Lisbon' and 'Villafranca' trees at Upland, California; was planted in the Rosenberger orchard and gained recognition as a superior cultivar. Tree closely resembles that of 'Villafranca'. Fruit is somewhat like 'Lisbon' but is shorter and broader and less tapered at base. Tree vigorous and prolific. Became popular in California in the 1960's.

'Rough Lemon' ('Florida Rough'; French'; 'Mazoe'; Jamberi')–perhaps a lemon X citron hybrid, but has been given the botanical name of C. *jambhiri* Lush. Believed to havc originated in northern India, where it grows wild; carried in 1498 or later by Portuguese explorers to southeastern Africa where it became naturalized along the Mazoe River; soon taken to Europe, and brought by Spaniards to the New World; is naturalized in the West Indies and Florida; oblate, rounded or oval, base flat to distinctly necked, apex rounded with a more or less sunken nipple; of medium size, averaging 2 3/4 in,(7 cm) wide, 2 1/2 (6.25 cm) high; peel lemon-yellow to orange-yellow, rough and irregular, with large oil glands, often ribbed; 3/16 to 3/8 in (5-10 mm) thick; pulp lemon-yellow, usually in 10 segments, medium-juicy, medium-acid, with moderate lemon odour and flavour; seeds small, 10 to 15, brownish within. Reproduces true from seeds, which are 96% to 100% nucellar. Tree large, very thorny; new growth slightly tinged with red; buds and flowers with red-purple.

The scant pulp and juice limit the rough lemon to home use. It is appreciated as a dooryard fruit tree in Hawaii and in other tropical and subtropical areas where better lemons are not available. The tree has been of great importance as a rootstock for the sweet orange, mandarin orange and grapefruit. It is not now used as a rootstock for lemon in Florida because of its susceptibility to "blight" (young tree decline). It is also prone to Alternaria leaf spot (*A1ternaria citri)* in the nursery, to foot rot *(Phytophthora parasitica).* Incidence varies with the clone and certain clones show significant resistance. In trials at Lake Alfred, 3 atypical clones showed immunity to leaf spot, while a typical rough lemon clone, 'Nelspruit 15', from South African seed, proved highly resistant to leaf spot and also extremely cold tolerant.

'Santa Teresa'–an old tree discovered to be disease-free in a 'Fermminello Ovale' orchard in Italy that had been devastated by *mal secco.* Budded trees from the original specimen were being commonly planted in the 1960's wherever the disease was prevalent in Italy.

Sweet Lemon (C. *limetta Risso)–a* general name for certain non-acid lemons or limettas, favoured in the Mediterranean region, In India, they are grown in the Nilgiris, Malabar and other areas. The fruits are usually insipid, occasionally subacid or acid. The seeds are white within and the tree is large, resembling that of the orange. One cultivar, called 'Dorshapo' after the plant explorers, Dorsett, Shamel and Popenoe, who introduced it from Brazil in 1914, resembles the 'Eureka' in most respects except for the lack of acidity. Another, called 'Millsweet', apparently was introduced into California from Mexico and planted in a mission garden. It was reproduced at the old University of California Experiment Station at Pomona. Neither is of any commercial value.

'Villafranca'–believed to have originated in Sicily; introduced into Sanford, Florida, from Europe around 1875 and later into California. Closely resembles 'Eureka'; of medium size. Tree is more vigorous, larger, more densely foliaged, and more thorny than 'Eureka' but becomes thornless with age. One strain is everbearing; another fruits heavily in summer. This was the leading lemon cultivar in Florida for many years; is cultivated commercially in Israel; is low-yielding and short-lived in India. It is little grown in California but has given rise to certain selections that are of importance, particularly 'Galligan Lisbon' and 'Corona Foothill Eureka'.

Climate: Because of its more or less continuous state of growth, the lemon is more sensitive to cold than the orange and less able to recover from cold injury. The tree is defoliated at 22º to 24º F (-5.56º-4.44º C). A temperature drop to 20º F (-6.67º C) will severely damage the wood unless there has been a fortnight of near-freezing weather to slow down growth. Flowers and young fruits are killed by 29º F (-1.67º C) and nearly mature fruits are badly damaged below 28º F (-2.22º C). On the other hand, the lemon attains best quality in coastal areas with summers too cool for proper ripening of oranges and grapefruit. Therefore, the lemon has a relatively limited climatic range. In Florida, lemons are produced commercially as far north as Ft. Pierce on the East Coast and Ruskin on the West Coast. The 'Meyer' lemon, as a dooryard tree, can be grown wherever oranges thrive, even as far west as Pensacola.

The fruits are scarred and the tree readily defoliated by winds, and benefit by the protection of windbreaks. Lemons are grown in both dry and humid atmospheres, the latter being a disadvantage mainly in the processes of curing and storing. Over a large lemon-growing region in California, annual rainfall varies from 25 to 125 cm. In long, dry periods, the lemon must be irrigated.

Soil: The lemon tree has the reputation of tolerating very infertile, very poor soil. In Florida, groves are mostly on sand. In California, excellent growth is maintained on silty clay loam of high water-holding capacity. In Guatemala, recommended soils are sand, clay and sandy-clay-deep, with high permeability and good drainage. Black soils are also suitable if not lying over calcareous subsoil. Ph should be between 5.5 and 6.5. If acidity is high, it is necessary to apply lime to achieve the optimum level.

Propagation: The rough lemon is widely grown from seed. The 'Meyer' lemon is easily reproduced by rooting large cuttings in the nursery and planting them directly in the grove. They fruit 2 to 3 years sooner than budded trees and have a long life, remaining in full production for over 30 years, perhaps much longer.

In Florida, commercial lemons have been budded onto 'rough lemon', sweet orange, and 'Cleopatra' mandarin rootstocks. More recent practices are the utilization of sour orange, Volkamer lemon (C. *volkameriana),* and alemow (C. *macrophylla* Wester, an old Philippine lemon/pummelo hybrid). The latter is employed in California on soils containing an excess of soluble salts and boron. If citranges are used as rootstocks for 'Eureka', bud union crease will kill the tree.

Culture: Lemon trees should be spaced 25 ft. (7.6 in) apart each way. If crowded or "hedged", production declines. The trees must be pruned when young and kept below 10 or 12 ft. (3-3.6 m) in height. They are cut back severely after 12 years or replaced. Weeds must be controlled but lemon trees are very sensitive to herbicides.

In Florida, fertilizing may be done 3 times a year between mid-November and the end of April, at the gradually increasing rate of 4 to 10 lbs (1.8-4.5 kg) per tree up to an age of 50 years. Nitrogen and potash are given in equal amounts under normal soil conditions. A nutritional spray with copper added is applied after spring bloom. Fertilizer and irrigation programmes should be varied according to the desired goal: fresh fruit marketing or processing. High nitrogen steps up yield and peel oil content but also results in more scab infection and poor curing. Potash increases acidity. Heavy irrigation increases yield and peel oil, scab infection, size of fruit and accelerates maturity.

In California, foliar spraying of urea is preferred over ground application of nitrogen which can lead to accumulations of salts and also contamination of groundwater. Leaf analyses are made to determine the nitrogen requirements of each cultivar for maximum yield. 'Eureka', in a 6-year test, showed no response to increased levels of nitrogen. In New Zealand, mature trees (15 to 20 years old) are given 25 to 30 lbs (11.3-13.6 kg) of complete mixed fertilizer annually, also heavy dressings of organic manure or mulch.

In Sicily, growers have, for over 50 years, made a practice of withholding water in summer-for 35 to 60 days-until the trees begin to wilt. Then the trees are heavily irrigated and given high nitrogen fertilizer which induces a second bloom in August or early September, producing a crop the following summer when lemons are scarce and prices are high. This system, called the "Verdelli process", was adopted on a little over 1,000 acres (405 ha) in California in 1983. Adequate bloom did not occur on sandy or shallow soils, but 80% of the plantings on gradually dehydrating, fine-textured soil bloomed well. Nearly $3 million was expected from this extra crop of summer lemons in the Central Valley and the Riverside area in 1984. New horticultural techniques are needed to overcome the handicaps of higher use of fertilizer, increased insect and fungus problems, effects of moisture stress on fruit quality, and low temperature hazard to immature fruits in winter.

In 1965, a team of California horticulturists initiated experimental trellis culture of 'Prior Lisbon' lemon on C. *macrophylla* rootstock. It was found that the labour of training, and repeated pruning either manually or by machine hedging and topping, was excessive and uneconomic. Guatemalan and Mexican growers interplant short-term crops such as beans, cassava, yautia*(Xanthosoma)*, in the rainy season, and tomatoes and peppers during the winter when the lemon trees will be irrigated and fertilized.

Harvesting and Handling: The marketability of lemons depends on the stage at which they are picked. Italian lemons for export are harvested as early as possible and are naturally "cured" in transit. In early days, California and Florida lemons were allowed to remain on the trees until they became too large. It was realized that early picking is necessary and California and Arizona adopted the practices of picking at any time after the fruits reach a 25 % juice content, and using rings to gauge the commercially acceptable size, and repeated spot-picking with clippers. Mechanical picking is impossible with lemons. The fruits are highly prone to oil spotting (oleocellosis) and cannot be handled roughly nor picked wet.

Formerly, Florida lemons were picked from mid-July to October for shipping fresh, and the balance in November was harvested for processing. Lemons under 2 1/8 in (5.4 cm) are too immature to attain proper quality for marketing and fruits over 2 1/2 in (6.25 cm) are too large. Manual spot-picking has been commonly practiced, but some producers have found it too costly, and are harvesting the entire crop at one time and grading for fresh sale or processing in the packing-house, discarding all undersized fruits. The lemons, after sorting according to colour, washing and coating with a fungicide and a thin layer of wax are stored (cured) until ready for shipping.

Yield: Lemon tree yields vary considerably with the cultivar, the location and weather conditions. A yield of 3 boxes per tree is commercially satisfactory in Florida. In India, a 6-year-old tree bore 966 fruits and, at 9 years of age, had produced a total of 3,173 fruits.

Storage: Florida's climate is unfavourable for long-term curing. It has been claimed that a 10-day curing period is adequate and degreening of Florida fruit is not needed. A major producer keeps the newly harvested fruits for 48 hours at 60º F (15.56º C) and 95% humidity, then passes them through a pre-grading procedure to eliminate all that are unusable. The usable fruits are then treated with fungicide against stem-end rot and returned to the curing room. Those harvested early in the season need 3 weeks to colour-up, the last may require less than a week. Finally, the fruits are washed, given a second fungicidal treatment, dried, waxed and packed.

Generally, lemons are cured at 56º to 58° F (13.33º-14.4º C) and 85-90% relative humidity. Green fruits may be held for 4 months or more, while the peel becomes yellow and thinner, the pulp juicier (6-80%) and the proportion of soluble solids higher (7-24%). Sometimes the degreening process is hastened by exposing the fruit to ethylene gas, ethephon, or silane, but this practice tends to stimulate decay, mainly through the shedding of the "button" (stem stub), the absence of which allows entry of *Diplodia natalensis, Phomopsis citri,* or *Alternaria* mycelium. Various auxins have been studied to determine which can be applied before storage to prevent button loss without delaying degreening. In 1982, Israeli investigators reported that decay losses from degreening procedures can be greatly reduced (from over 50% to 6.3%) by packaging the fruits in 10 micrometer-thick high-density polyethylene. This treatment makes it possible to store lemons with minimum damage for as long as 6 months.

In the past, New Zealand lemons for storage have been individually wrapped in diphenyl-treated paper after washing and dipping in a 200 ppm solution of 2,4,5-T and then waxing. The fruits were marketable after storing for 4 months at room temperature. Lemons can be kept for weeks in the home refrigerator if placed in a jar with a tight-fitting lid to prevent loss of moisture.

Lemons for export from Florida to Hawaii and Arizona must be fumigated with methyl bromide because of possible infestation by the Caribbean fruit fly. For sale within the state, other methods must be employed.

Pests and Diseases: In Southeast Asia, many species of ants attack the root system and the farmer times the opening of the water gates so as to force the ants to the surface of the beds, where he burns them with fire.

One of the 3 most serious arthropod pests of the lemon and other citrus trees in California is California red scale, *Aonidiella aurantii.* In the southern part of the state it is under biological control but it requires applications of pesticides in the San Joaquin Valley. In Florida, rust mites, purple mites and purple scale may at times be troublesome but they are all controllable with appropriate sprays.

Young lemon trees in California sometimes require protection from wild rabbits.

Diseases are the greater challenges. In Florida, the main lemon diseases are scab *(Elsinoe fawcetti)* on fruit, leaves and twigs; anthracnose of fruit (stylar-end-rot), leaves and twigs caused by both *Colletotrichum gloeosporioides* and *Glomerella cingulata;* greasy spot *(Mycosphaerella citri* or *Cercospora citri-grisea);* and gummosis *(Diaporthe citri).* The latter organism also causes melanose and die-back, and stem-end rot. Stem-end rot may also arise from attack by *Botryosphaeria ribis* and *Diplodia natalensis.*

Other lemon diseases recorded in Florida are branch knot *(Sphaeropsis tumefaciens),* damping-off *(Rhizoctonia solani),* leaf spot *(Mycosphaerella horii, Alternaria citri,* and *Catenularia sp.;* algal leaf spot or green scurf *(Cephaleuros virescens);* tar spot *(Cercospora gigantea);* felt fungus *(Septobasidium pseudopedicellatum);* charcoal root rot *(Macrophomia phaseolina);* root rot *(Fusarium oxysporum, Pythium ultimum,* and *Phytophthora parasitica;* heart rot and wood rot *(Fomes applanatus* and *Ganoderma sessilis);* crinkly leaf and exocortis viruses; and green mold *(Penicillium digitatum);* blue mold (P. *italicum);* and pink mold (P. *roseum).* In 1955, the lemon budwood certification programme was begun to provide virus-free stock for growers.

Red algae infests lemon trees and causes much dieback unless controlled with copper fungicide in the summer. Zinc deficiency causes stunting of twigs, reduced flowering, premature dropping of fruit, and yellow bands along the leaf veins. Manganese deficiency is evidenced by interveinal chlorosis and subsequent necrosis, shedding of leaves, flowers and young fruit. In India, fruit cracking occurs when dry periods are followed by heavy rains. Cracking can be largely avoided by frequent light irrigation during the dry period and early picking.

Stored lemons are subject to the stem-end rots and the molds listed above. The albedo may show small dark sunken areas even though this defect is not visible externally. Cultivars differ in their ability to resist decay.

Food Uses: Slices of lemon are served as a garnish on fish or meat or with iced or hot tea, to be squeezed for the flavourful juice. In Colombia, lemon soup is made by adding slices of lemon to dry bread roll that has been sauteed in shortening until soft and then sieved. Sugar and a cup of wine are added and the mixture brought to a boil, and then served.

Lemon juice, fresh, canned, concentrated and frozen, or dehydrated and powdered, is primarily used for lemonade, in carbonated beverages, or other drinks. It is also used for making pies and tarts, as a flavoring for cakes, cookies, cake icings, puddings, sherbet, confectionery, preserves and pharmaceutical products. A few drops of lemon juice, added to cream before whipping, gives stability to the whipped cream. Lemon peel can be candied at home and is preserved in brine and supplied to manufacturers of confectionery and baked goods. It is the source of lemon oil, pectin and citric acid. Lemon oil, often with terpenes and sesquiterpenes removed, is added to frozen or otherwise processed lemon juice to enrich the flavour. It is much employed as a flavoring for hard candies.

Toxicity: The thorns of the lemon tree inflict painful punctures and scratches. Lemon peel oil may cause contact dermatitis, chronic in those who handle, cut and squeeze lemons daily. Parts of the body touched by contaminated hands may show severe reactions after exposure to the sun. People that suck lemons may suffer irritation and eruptions around the mouth. The wood of lemon trees and its saw-dust may induce skin reactions in sensitive woodworkers.

OTHER USES

Lemon juice is valued in the home as a stain remover, and a slice of lemon dipped in salt can be used to clean copper-bottomed cooking pots. Lemon juice has been used for bleaching freckles and is incorporated into some facial cleansing creams. Lemon peel oil is much used in furniture polishes, detergents, soaps and shampoos. It is important in perfume blending and especially in colognes. Petitgrain oil (up to 50% citral), is distilled from the leaves, twigs and immature fruits of the lemon tree in West Africa, North Africa and Italy. With terpenes removed, it is greatly prized in colognes and floral perfumes.

Lemon peel, dehydrated, is marketed as cattlefeed. Lemonade, when applied to potted plants, has been found to keep their flowers fresh longer than normal. But it cannot be used on chrysanthemums without turning their leaves brown.

Wood: The wood is fine-grained, compact, and easy to work. In Mexico, it is carved into chessmen, toys, small spoons, and other articles.

Medicinal Uses: Lemon juice is widely known as a diuretic, antiscorbutic, astringent, and febrifuge. In Italy, the sweetened juice is given to relieve gingivitis, stomatitis, and inflammation of the tongue. Lemon juice in hot water has been widely advocated as a daily laxative and preventive of the common cold, but daily doses have been found to erode the enamel of the teeth. Prolonged use will reduce the teeth to the level of the gums. Lemon juice and honey, or lemon juice with salt or ginger, is taken when needed as a cold remedy. It was the juice of the Mediterranean sweet lemon, not the lime, that was carried aboard British sailing ships of the 18th Century to prevent scurvy, though the sailors became known as "limeys".

Oil expressed from lemon seeds is employed medicinally. The root decoction is taken as a treatment for fever in Cuba; for gonorrhea in West Africa. An infusion of the bark or of the peel of the fruit is given to relieve colic.

6

Soil, Plant and Water

SOILS

Sampling: Soil sampling is a particularly difficult task when attempting to get a representative sample. Normally a 500-gram sample is submitted to the laboratory for analysis. This 500-gram sample may represent 10 or more acres. If the area covered by the sample is not uniform, the chemical analysis may not accurately reflect the nutrient status of specific sites. Factors that need to be considered when sampling soil include the depth and time of sampling. Proper sampling depth is affected by the crop being grown, past cropping, depth of plowing and also the nutrient of interest. Subsoil samples are important for most crops. Standard sampling times should be used due to the difficulty in comparing samples taken at different times. The fertility level of a field will vary over the course of the year and interpreting results for samples taken at different times of the year will be very difficult. Sampling between crops will give more consistent results.

When sampling soils, the area should be subdivided into as homogeneous sections as possible. Between 10 and 20 sub-samples should be composited from each area. Sub-samples should be small enough that the composite sample will be of a size that can be completely processed for analysis. The depth of the sampling is determined by the crop, the elements of interest and existing knowledge about the soil profile. Samples for cultivated crops are taken from the plow layer. Pasture and sod crop soil samples should be collected from the top four inches. Samples for nitrate, soluble salts and available micronutrients should be taken at the root depth.

Drying Recommendations: Once samples have been collected, they must be processed promptly to prevent any changes that might affect the analysis. Break up large chunks of soil and spread out to air dry where the sample will not be contaminated, particularly by fertilizer dust. The sample may also be placed in a forced air oven set between 35°C and 55°C.

Grinding Recommendations: Soil should be crushed in a soil pulverizer to pass through a 20 mesh screen. Large clumps of hard soil should be pounded into smaller clumps first. Soil pulverizing time is normally one to four minutes per sample.

Exceptions:

- Analysis of Carbon, Nitrogen (including Total Kjedahl Nitrogen), and total elements (Arsenic, Cadmium, Copper, Chromium, Iron, Lead, Manganese, Molybdenum, Nickel, Phosphorus, Selenium, Zinc) requires that the sample be ground to a powder fine enough to pass through a 60 mesh screen.
- Soils with a high percentage of sand or organic material may require grinding to pass through a 60 mesh screen to achieve sample homogeneity.

PLANTS

Sampling: Sampling plays a critical role in plant analysis. When analyzing the nutrient status of plants, it is essential to select the plant part for chemical analysis that reflects the status of the particular element of interest. Four samplings during a growing season are usually sufficient to characterize seasonal nutritional patterns.

One sampling should be early in the growing season, two in mid-season and the last one just prior to harvest. Four samples should be collected from each field or management unit. Each sample should contain material from at least 20 plants to ensure adequate, representative material for analytical testing. Separate samples should be taken from areas that appear different from the rest of the field.

A young mature leaf is generally selected for analysis. The sample can be subdivided into blade and petiole. The status of Cl, NO_3-N, NH_4-N, extractable K and P, in the form of PO_4-P (2% acetic acid) are generally determined through analysis of the petiole. Blades are used when evaluating the status of K, Ca, Mg, Na, Fe, Mn, Zn, Cu, B, Mo, SO_4-S and total-N in plants. For diagnostic purposes, only leaves that have recently developed symptoms should be collected for chemical analysis.

Drying Recommendations: After collection, plant material should be washed to remove any residual soil or dust. Fresh samples or those suspected to be moist should be placed into paper bags (with adequate room for air movement within the bag) and dried in a forced air oven at 55-60°C. In general, adequate drying time is approximately 12 hours or until the material snaps or breaks easily. All samples, except freeze drying material, should be turned every 24 hours.

Exceptions:

- Range samples take 24 hours or more to dry
- Power plant ash samples take three days to dry
- Solid pieces of wood take 24 hours to dry
- Rice soils take two days to dry
- Any material larger than gallon size will take one or more days to dry
- Fruits and vegetables should be freeze dried and pureed (weigh the samples before and after the freeze drying process).

These time frames are based on use of a large, forced-air oven. Times need to be adjusted accordingly for other types/sizes of ovens.

Grinding Recommendations: Most plant and feed samples should be ground to pass through a 40 mesh screen. Large wood samples must be splintered into smaller pieces before grinding. Average grinding time is one minute per sample if less than 10 grams or three to four minutes for samples over 10 grams.

Exceptions:

- Grape blades-use a 20 mesh screen due to tricomb separation from leaf
- Sugar beet petioles-use a 20 mesh screen due to high sugar content
- Walnut leaf-use a 20 mesh screen due to fibers in veins
- Corn stalks-use a 10 mesh screen due to high sugar content
- Freeze dried samples-do not need to be ground as they are easily crushed manually.

WATER

Sampling: Samples must be representative. Samples should be collected in clean, plastic bottles that have been rinsed three times prior to use. Well-water samples should be collected after pumping for at least 30 minutes. Sampling from distribution systems should be done after the lines have been flushed sufficiently to ensure that the sample is representative of the supply.

If NO_3-N, NH_4-N or PO_4-P are elements of interest the samples should be frozen or kept below 40° F. Samples collected for alkalinity, conductivity, phosphate, sulfate, turbidity or solids should also be refrigerated until analysis can be completed. Due to the problems of absorption or precipitation, if micronutrients, metals or salts are of interest, the sample, or a sub-sample, should be acidified to pH <2 as soon as possible after collection. Freezing is a good choice for water sample preservation in many cases. Samples high in carbonates should not be frozen.

Please take the following precautions when sending frozen water samples:

- Always leave at least 10% of the container unfilled to accommodate the expansion of the water when frozen.
- Do not place sample containers directly onto dry ice. Place some insulation between the dry ice and the sample containers.
- Do not use glass bottles the sheer bulk of the sample becomes significant. Soil samples of over 50 mg are very difficult to analyse due to rapid ash buildup in the furnace.

Combustion capsule formation is critical for successful analytical runs. Please refer to the detailed instructions for further information.

SETTING THE STANDARD IN SOIL SAMPLING

New standards when sampling for volatiles in soil will demand new processes and new equipment according to Vincent Van Walt.

If you are involved in sampling for volatile compounds in soils – things have changed with new regulations and standards coming into force in the USA (USA-EPA Guideline 5035a_r1) and with similar in preparation in Europe through the ISO. In summary these new standards recommend that less soil (16ml ~ 25g) is taken for laboratory analysis and that any soil to be investigated should be sampled with a limited amount of disruption during the extraction process in order to minimise VOC losses prior to sample preparation and determination in the laboratory. This usually means collecting the soil sample with a single transfer to an air tight container that will be used for storage and preparation for analysis. In addition, acceptable sampling devices, procedures, preservatives and techniques are required to be recorded and observed.

Previously samples of soils for the analysis of very volatile components such as benzene, toluene and chlorinated hydrocarbons were normally obtained directly from a borehole. The stainless steel sample tubes containing the soil were closed completely and cooled for further transport to the laboratory for analysis. The new regulations recommend that a small, accurately measured sub-sample be taken above ground from a larger sampler or auger.

For environmentalists in the field, this new closed-system purge-and-trap process for the analysis of volatile compounds in solid materials like soils, sediments, and solid waste requires different equipment. For a low soil method a hermetically-sealed sample vial is required, the seal of which is never broken from the time of sampling to the time of analysis and since the sample is never exposed to the atmosphere after sampling, the losses of VOCs during sample transport, handling, and analysis should be minimised. The applicable concentration range of the low soil method is dependent on the determinative method, matrix, and compound. However, it will generally fall in the 0.5 to 200ìg/kg range.

As with any preparative method for volatiles, samples should be screened to avoid contamination of the purge-and-trap system by samples that may contain very high concentrations of purgeable materials. In addition, because the sealed sample container cannot be opened to remove a sample without compromising the integrity of the sample, multiple samples should be collected to allow for screening and reanalysis.

However, not all closed-system purge-and-trap equipment employed for low concentration samples is appropriate for soil samples preserved in the field with methanol. Precautions should be taken when preserving a soil by this method since certain compounds within the olefins, ketones, esters, ethers, and sulfides classes may react under low pH conditions and possibly not be representative of the material sampled. Additionally, acidification of solid wastes may evolve toxic gases that may be harmful to field and laboratory personnel. It is therefore recommended that when collecting wastes of unknown composition, preliminary screening and characterisation of potential sample contents should be performed prior to use of acidification as a means to chemically preserve samples designated for determinative analyses.

After a fresh surface of the solid material to be analysed is exposed to the atmosphere, the sub-sample collection process should be completed as quickly as possible in order to minimise the loss of VOCs due to volatilisation. Removing the sub-sample from a material should be done with the least amount of disruption (disaggregation) as possible. Additionally, rough trimming of the sampling location's surface layers should be considered if the material may have already lost VOCs (been exposed for more than a couple of minutes) or if it may be contaminated by other waste, different soil strata, or vegetation. Removal of surface layers can be accomplished by scraping the surface using a clean spatula, scoop, knife, or shovel.

Sub-sampling should be carried out using an appropriate device. The new standards recommend a smaller coring device that will help to maintain the sample structure during collection and transfer as do their larger counterparts used to retrieve subsurface materials. When inserting a clean coring tool into a fresh surface for sample collection, air should not be trapped behind the sample. If air is trapped, it could either pass through the sampled material causing VOCs to be lost or cause the sample to be pushed prematurely from the coring tool.

After an undisturbed sample has been obtained by pushing the barrel of the coring tool into a freshly exposed surface and then removing the corer once filled, the exterior of the barrel should be quickly wiped with a clean disposable towel. The next step varies, depending on whether the coring device is used for sample storage and transfer or solely for transfer. If the coring tool is used as a storage container, cap the open end after ensuring that the sealing surfaces are cleaned. If the device is to be solely used for collection and not for storage, immediately extrude the sample into a VOA vial or bottle. The volume of material collected should not cause excessive stress on the coring tool during intrusion into the material, or be so large that the sample easily falls apart during extrusion.

Obtaining and transferring a sample should be done rapidly (<10 seconds) to reduce volatilisation losses. If the vial or bottle contains methanol or another liquid, it should be held at an angle when extruding the sample into the container to minimise splashing. Just before capping, a visual inspection of the lip and threads of the sample vessel should be made, and any foreign debris should be removed with a clean towel, allowing an airtight seal to form.

Following the new standards, a site-specific Sampling and Analysis Plan should clearly list the required sample collection equipment necessary to ensure that the loss of volatile constituents is minimised during the sample collection process. As with all environmental sampling applications, the analytical data usability and representativeness will be affected by improper sample collection techniques.

Sampling personnel will be responsible for ensuring that VOA vials are sealed properly using a septum of sufficient thickness without any punctures. The improper vial sealing (i.e. due to excess sample retained on the vial threads) and tightening of caps are the primary factors in the loss of volatiles due to sample collection activities. Care should also be exercised in the selection of approved pre-cleaned and certified VOA vials absent of burrs on the glass.

Procedures should be in place for the selection and appropriate use of sample collection devices (i.e. bailer, coring tool, etc.) along with the required decontamination measures. It is also recommended to store a control sample when collecting volatile samples in order to assess possible field induced contamination.

Whichever method is most appropriate for the volatile and soil conditions to be sampled the equipment chosen should be easy to decontaminate so a high grade stainless steel is ideal. Aluminium lined seals will prevent sorption and diffusion and, the mechanism of the corer should aim to eliminate headspace and therefore loss of VOCs from the soil.

The new standards will improve the quality of results obtained when sampling for volatiles in soil however the problem of obtaining the sample should not be understated. The problem lays in the field not the laboratory so the soil sampler used, the time of exposure prior to sub-sampling, wind conditions, temperature and the type of volatile being tested for will all impact on the potential losses.

There is no doubt with the new standards preservation of the volatiles is optimised and freezing of the sample mitigates volatilisation but it is still the first steps in the process that remain the Achilles heel-but who am I to argue with the esteemed authors of guidelines or standards?

THE LIVING SOIL: TEXTURE AND STRUCTURE

Soils are made up of four components: minerals, air, water, and organic matter. In most soils minerals represent around 45% of the total volume, water and air about 25% each, and organic matter from 2% to 5%. The mineral portion consists of three distinct particle sizes classified as sand, silt or clay. Sand is the largest size particle that can be considered soil. Sand is largely the mineral quartz, though other minerals are also present. Since quartz contains no plant nutrients, sand is the lowest contributor to soil fertility of the three soil particle sizes.

Furthermore, sand cannot hold nutrients–they leach out easily with rainfall. That is why sandy soils are not as productive as loams and need to be spoon-fed fertilizer. Silt particles are much smaller than sand but, like sand, silt is mostly quartz.

The smallest of all the soil particles is clay. Clays are quite different from sand or silt and contain appreciable amounts of plant nutrients. Clay has a large surface area resulting from the plate-like shape of the individual particles. The textural designation of a soil is derived from the relative portions of sand, silt, and clay. A sandy loam, for example, has much more sand and much less clay than does a clay loam. A loam soil is a mixture of sand, silt and clay. Most soils are some type of loam. They are more accurately described by the words the preface the word loam, such as: sandy loam or clay loam. The texture designations are found in Table.

Another soil characteristic–soil structure–is different from soil texture. Structure refers to the combination or "aggregation" of sand, silt and clay particles into larger secondary clusters. If you grab a handful of soil, good structure is apparent when the sand, silt, and clay particles are aggregated into granules or crumbs. Both texture and structure determine pore space for air and water circulation, erosion resistance, looseness, ease of tillage, and root penetration. However, while texture is an innate property of the native soil and does not change with agricultural activities, structure can be improved or destroyed readily through our choice and timing of farm practices. The organic soil component contains all the living creatures in the soil and the dead ones in various stages of decomposition. An acre of living soil can contain 900 pounds of earthworms, 2400 pounds of fungi, 1500 pounds of bacteria, 133 pounds of protozoa, 890 pounds of arthropods and algae, and even small mammals in some cases. In fact, the soil could be viewed as a living entity, rather than an inert body.

The soil's organic matter also contains dead organisms, plant matter and other organic materials in various phases of decomposition. Humus, the dark-coloured organic material in the final stages of decomposition, is relatively stable. Both organic matter and humus serve as a reservoir of plant nutrients; they also help to build soil structure and provide other benefits.

The type of healthy living soil required to support humans now and far into the future will be balanced in nutrients and high in humus with a high diversity of soil organisms. It will produce healthy plants with minimal weed, disease and insect pressure. To accomplish this we work with the natural processes and optimize their functions to sustain our farms.

THE LIVING SOIL: IMPORTANCE OF SOIL ORGANISMS

If you look out at a landscape you might wonder how native prairies and forests function in the complete absence of tillage and fertilizers? These soils are tilled by soil organisms, not by machinery. They are fertilized too, but the fertility is used again and again and never leaves the site. Native soils are covered with a layer of plant litter and/or growing plants throughout the year. Beneath the surface litter layer, a rich complexity of soil organisms decompose plant residue and dead roots, then release their stored nutrients slowly over time. In fact, topsoil is the most biologically diverse part of the earth. Soil-dwelling organisms release bound-up minerals converting them into plant-available forms that are then taken up by the plants growing on the site. The organisms recycle nutrients again and again from the death and decay of each new generation of plants growing on the site.

There are many different types of creatures that live on or in the soil. Each has a role to play. These organisms will work for the farmer's benefit if we simply manage for their survival. Consequently we may refer to them as soil livestock. While there is a great variety of organisms that contribute to soil fertility, earthworms, arthropods, and the various microorganisms merit particular attention.

Earthworms: Earthworm burrows enhance water infiltration and soil aeration. Earthworm tunnelling can increase the rate of water entry into the ground 4 to 10 times higher than fields that lack worm tunnels. This reduces water runoff, recharges groundwater, and helps store more soil water for dry spells. Vertical earthworm burrows pipe air deeper into the soil, stimulating microbial nutrient cycling at those deeper levels. Tillage done by earthworms can replace some expensive tillage work done by machinery.

Worms eat dead plant material left on top of the soil and redistribute the organic matter and nutrients throughout the topsoil layer. Nutrient-rich organic compounds line the tunnels that may remain in place for years if not disturbed. During droughts these tunnels allow for deep plant root penetration into subsoil regions of higher moisture content. In addition to organic matter, worms also consume soil and soil microbes as they move through the soil. The soil clusters they expel from their digestive tracts is known as a worm cast or casting. Each worm cast is separate from other casts and ranges in size from that of a mustard seed to a sorghum seed depending on the size of the worm. The soluble nutrient content of worm casts is considerably higher than those of the original soil. A good population of earthworms can process 20,000 pounds of topsoil per year, with turnover rates as high as 200 tons per acre having been reported in some exceptional cases.

Earthworms also secrete a plant growth stimulant. Reported increases in plant growth due to earthworm activity may be attributed to this substance, not just to improved soil quality.

Table : Nutrients in worm casts compared to the surrounding soil from which they came.

Nutrient	Worm casts	Soil
	Lbs/ac	Lbs/ac
Carbon	171,000	78,500
Nitrogen	10,720	7000
Phosphorus	280	40
Potassium	900	140

Earthworms thrive where there is no-tillage–generally, the less tillage, the better, and the shallower the tillage, the better. Worm numbers can be reduced by as much as 90% by deep and frequent tillage. Tillage reduces earthworm populations by drying the soil, burying the plant residue they feed on, and making the soil easier to freeze. Tillage also destroys their vertical burrows and can kill and cut up the worms themselves. Emergence times for young worms are spring and fall–their most active periods just when most farmers are interested in tillage. Worms are dormant in the hot part of the summer and the cold of winter. Table below shows the effect of tillage and cropping practices on earthworm numbers.

Table : Effect of crop management on earthworm populations.

Crop	*Management*	*Worms/foot2*
Corn	Plow	1
Corn	No-till	2
Soybean	Plow	6
Soybean	No-till	14
Bluegrass/clover	—	39
Dairy pasture	—	33

As a rule, earthworm numbers can be increased by reducing or eliminating tillage (especially fall tillage), never using the moldboard plow, reducing residue particle size (using a straw chopper on the combine), adding animal manure, and growing green manure crops. It is beneficial to leave as much surface residue as possible year round. Cropping systems that typically have the most earthworms are in descending order: perennial cool-season grass grazed rotationally, then warm-season perennial grass grazed rotationally, then annual croplands using no-till. Ridge-till and strip tillage will generally have more earthworms than clean tillage involving plowing and disking.

Earthworms prefer a near neutral soil pH, moist soil conditions, and plenty of plant residue on the soil surface. They are sensitive to certain pesticides and some incorporated fertilizers. Carbamate insecticides including Furadan, Sevin, and Temik, are harmful to earthworms, notes worm biologist Clive Edwards of Ohio State University. Some insecticides in the organophosphate family are mildly toxic to earthworms while synthetic pyrethroids are harmless to them. Most herbicides have little effect on worms except for the triazines, such as Atrazine, which are moderately toxic. Also, anhydrous ammonia kills earthworms in the injection zone.

Arthropods: In addition to earthworms, there are many other species of soil organisms that can be seen by the naked eye. Among them are sowbugs, millipedes, centipedes, slugs, snails and springtails. These are the primary decomposers.

Their role is to eat and shred the large particles of plant and animal residues. Some bury residue, bringing it into contact with other soil organisms that further decompose it. Some members of this group prey on smaller soil organisms. The springtails are a small insect, which eat mostly fungi. Their waste is rich in plant nutrients that are released after other fungi and bacteria decompose it. Also of interest are dung beetles, which play a valuable role in recycling manure and reducing livestock intestinal parasites and flies.

Bacteria: Most numerous among soil organisms are the bacteria; every gram of soil contains at least a million of these tiny one-celled organisms. There are many different species of bacteria, each with its own role in the soil environment.

One of the major benefits bacteria provide for plants is in helping them take up nutrients. Some species release nitrogen, sulfur, phosphorus, and trace elements from organic matter. Others break down soil minerals and release potassium, phosphorus, magnesium, calcium and iron. Still other species make and release natural plant growth hormones, which stimulate root growth.

A few species of bacteria fix nitrogen in the roots of legumes while others fix nitrogen independently of plant association. Bacteria are responsible for converting nitrogen from ammonium to nitrate and back again depending on certain soil conditions. Other benefits to plants provided by various species of bacteria include increasing the solubility of nutrients, improving soil structure, fighting root diseases, and detoxifying soil.

Fungi: Fungi come in many different species, sizes and shapes in soil. Some species appear as thread-like colonies, while others are one-celled yeasts. Slime molds and mushrooms are also fungi. Many fungi aid plants by breaking down organic matter or by releasing nutrients from soil minerals. Fungi are generally early to colonize larger pieces of organic matter and begin the decomposition process. Some fungi produce plant hormones, while others produce antibiotics including penicillin. There are even species of fungi that trap harmful plant-parasitic nematodes.

The mycorrhizae (my-cor-ry-'zee) group of fungi lives either on or in plant roots and act to extend the reach of root hairs into the soil. Mycorrhizae increase the uptake of water and nutrients especially in less fertile soils. Roots colonized by mycorrihizae are less likely to be penetrated by root-feeding nematodes since the pest cannot pierce the thick fungal network. Mycorrhizae also produce hormones and antibiotics, which enhance root growth and provide disease suppression. The fungi benefit from plant association by taking nutrients and carbohydrates from the plant roots they live in.

Actinomycetes: Actinomycetes are thread-like bacteria that look like fungi. While not as numerous as bacteria, they also perform vital roles in the soil. Like the bacteria, they help decompose organic matter into humus, releasing nutrients. They also produce antibiotics to fight diseases of roots. These same antibiotics are used to treat human diseases. Actinomycetes are responsible for the sweet, earthy smell of biologically active soil noticed whenever a field is tilled.

Algae: Many different species of algae also live in the upper half-inch of the soil. Unlike most other soil organisms, algae actually produce their own food through photosynthesis. They appear as a greenish film on the soil surface following a good rain. Algae improve soil structure by producing slimy substances that glue soil together into water-stable aggregates. Some species of algae (the blue-greens) can fix their own nitrogen, some of which is later released to plant roots.

Protozoa: Protozoa are free-living microorganisms that crawl or swim in the water between soil particles. Many soil protozoa are predatory, eating other microbes. One of the most common is an amoeba that eats bacteria. By eating and digesting bacteria, protozoa speed up the cycling of nitrogen from the bacteria, making it more available to plants.

Nematodes: While nematodes are abundant in most soils, only a few species are harmful to plants. The harmless species eat decaying plant litter, bacteria, fungi, algae, protozoa and other nematodes. Like other soil predators, nematodes speed the rate of nutrient cycling.

All these organisms–from the tiny bacteria up to the large earthworms and insects–interact with one another in a multitude of ways in a whole soil ecosystem. Organisms not directly involved in decomposing plant wastes may feed on each other or each other's waste products or the other substances they release. Among the other substances released by the various microbes are vitamins, amino acids, sugars, antibiotics, gums, and waxes.

Roots can also release various substances into the soil that stimulate soil microbes. These substances serve as food for select organisms. Some scientists and practitioners theorize that plants use this means to stimulate the specific population of microorganisms capable of releasing or otherwise producing the kind of nutrition needed by the plants.

Research on biological life in the soil has determined that there are ideal ratios for certain key soil organisms in highly productive soils (soil foodweb).

The Soil Foodweb lab, located in Oregon, tests soils and makes fertility recommendations that are based on this understanding. Their goal is to alter the makeup of the soil microbial community so it resembles that of a highly fertile and productive soil.

There are several different ways to accomplish this goal, depending on the situation. Because we cannot see most of the creatures living in the soil and may not take time to observe the ones we can see, it is easy to forget about them. See Table below for estimates of typical amounts of various organisms found in fertile soil. There are many websites that provide in-depth information on soil organisms. Look for a list of these web sites in the Additional Information Resources section of this publication. Many of these web sites have colour photographs of soil organisms and describe their benefits.

Table : Weights of soil organisms in the top 7 inches of fertile soil.

Organism	Pounds of liveweight/acre
Bacteria	1000
Actinomycetes	1000
Molds	2000
Algae	100
Protozoa	200
Nematodes	50
Insects	100
Worms	1000
Plant roots	2000

FERTILIZER AMENDMENTS AND BIOLOGICALLY ACTIVE SOILS

What are the soil mineral conditions that foster biologically active soils? Drawing from the work of Dr. William Albrecht (1888 to 1974), agronomist at the University of Missouri, we learn that balance is the key. Albrecht advocated bringing soil nutrients in balance so that none were in excess or deficient. Albrecht's theory (also called base-saturation theory) is used to guide lime and fertilizer application by measuring and evaluating the ratios of positively charged nutrients (bases) held in the soil. The positively charged bases include calcium, magnesium, potassium, sodium, ammonium nitrogen, and several trace minerals. When optimum ratios of bases exist, the soil is believed to support high biological activity, becomes resistant to leaching, and has optimal physical properties (water intake and aggregation). The plants growing on such a soil are also balanced in mineral levels and are nutritious to humans and animals alike.

Through extensive research, Albrecht determined the desirable percentages of base saturation in the soil. These percentages, he maintained, were optimal for the growth of most crops.

These levels are:

Calcium	60-70%
Magnesium	10-20%
Potassium	2-5%
Sodium	0.5-3%
Other bases	5%

Fertilizer and lime applications should be made at rates that will bring soil mineral percentages into this ideal range. Through this approach, soil pH shifts automatically into a desirable range without creating nutrient imbalances. The base saturation theory also takes into account the effect one nutrient may have on another and avoids undesirable interactions. For example, excess phosphorus is known to tie up zinc.

The Albrecht system of soil evaluation contrasts with the approach used by many state laboratories often called the "sufficiency method." Sufficiency theory places little to no value on nutrient ratios, and lime recommendations are typically based on pH measurements alone. While in many circumstances base saturation and sufficiency methods will produce identical soil recommendations and similar results, significant differences can occur on a number of soils. For example, suppose we tested a cornfield and found a soil pH of 5.5 and base saturation for magnesium at 20% and calcium at 40%. Base saturation theory would call for liming with a high-calcium lime to raise the % base saturation of calcium; the pH would rise accordingly. Sufficiency theory would not specify high calcium lime and the grower might choose instead, a high magnesium dolomite lime that would raise the pH but worsen the balance of nutrients in the soil. Another way to look at these two theories is that the base saturation theory does not concern itself with pH to any great extent but rather with the proportional amounts of bases. The pH will be correct when the levels of bases are correct.

Albrecht's ideas have found their way onto large numbers of American farms and into the programs of several agricultural consulting companies. Neal Kinsey, a soil fertility consultant of Charleston, MO, is a major proponent of the Albrecht approach. Kinsey was a student under Albrecht and is one of the leading authorities on the base-saturation method. He teaches a short course on the Albrecht system and provides a soil analysis service. His book, *Hands On Agronomy*, is widely recognized as a highly practical guide to its understanding and implementation.

Several firms–many providing backup fertilizer and amendment products–offer a biological-farming program based on the Albrecht theory. Typically these firms offer broad-based soil analysis and recommend balanced fertilizer materials considered friendly to soil organisms. They avoid the use of some common fertilizers and amendments such as dolomite lime, potassium chloride, anhydrous ammonia, and oxide forms of trace elements because they are considered harmful to soil life.

CONVENTIONAL FERTILIZERS

Commercial fertilizer can be a valuable resource to farmers in transition to a more sustainable system and can help meet nutrient needs during times of high crop nutrient demand, or when weather conditions result in slow nutrient release from organic resources. Commercial fertilizers have the advantage of supplying plants with immediately available forms of nutrients. They are often less expensive and less bulky to apply than processed natural fertilizers.

Not all conventional fertilizers are alike, however. Many appear harmless to soil livestock but a few are problematic. Anhydrous ammonia contains approximately 82% nitrogen and is applied subsurface as a gas. Anhydrous speeds the decomposition of organic matter in the soil, leaving a soil more compact as a result. The addition of anhydrous contributes acidity to the soil, requiring 148 pounds of lime to neutralize 100 pounds of anhydrous ammonia or 1.8 pounds of lime for every pound of nitrogen contained in the anhydrous. Anhydrous ammonia initially kills many soil microorganisms in the application zone. Bacteria and actinomycetes recover within one to two weeks to levels higher than those prior to treatment. Soil fungi, however, may take seven weeks to recover. During the recovery time, the bacteria are stimulated to grow and decompose more organic matter due to the high soil nitrogen content. This is why their numbers increase after anhydrous applications. Farmers commonly report that the long-term use of synthetic fertilizers, especially anhydrous ammonia, leads to soil compaction and poor tilth. When bacteria increase and organic matter decreases, aggregation naturally declines because there is no more glue being produced to stick the soil particles together.

Potassium chloride (KCl) (0-0-60 and 0-0-50), also known as muriate of potash, contains approximately 50 or 60% potassium and 47.5% chloride.

Muriate of potash is made by refining potassium chloride ore, which is a mixture of potassium and sodium salts and clay from the brines of dying lakes and seas. The potential harmful effects from KCl can be surmised from in the salt concentration of the material. Table below shows that, pound for pound, KCl is surpassed only by table salt on the salt index. Additionally, some plants such as tobacco, potatoes, peaches and some legumes are especially sensitive to chloride. High rates of KCl must be avoided on such crops. Potassium sulfate, potassium nitrate, sul-po-mag, or organic sources of potassium may be considered as alternatives to KCl for fertilization.

Sodium nitrate, also known as Chilean nitrate, or nitrate of soda, is another high salt fertilizer. Because of the relatively low nitrogen content of sodium nitrate, a high amount of sodium is added to the soil when normal applications of nitrogen are made with this material.

The concern is that excessive sodium acts as a dispersant of soil particles, degrading aggregation. The salt index for KCl and sodium nitrate can be seen in Table below.

Table : Salt index for various fertilizers.

Material	Salt Index	Salt index per unit of plant food
Sodium Chloride	153	2.9
Potassium chloride	116	1.9
Ammonium nitrate	105	3.0
Sodium nitrate	100	6.1
Urea	75	1.6
Potassium nitrate	74	1.6
Ammonium sulfate	69	3.3
Calcium nitrate	53	4.4
Anhydrous ammonia	47	.06
Sulfate-potash-magnesia	43	2.0
Di-ammonium phosphate	34	1.6
Monammonium phosphate	30	2.5
Gypsum	8	.03
Calcium carbonate	5	.01

BIOLOGICAL AND CHEMICAL ASPECTS OF SOIL PRODUCTIVITY

Observations such as the accumulative beneficial effects of pasture, and the degradative impacts of successive cereal crops on a coarse-textured sandy loam soil in Australia provide a basis for recommending rotations to maintain sustainability. There are many beneficial effects from growing legume crops and pastures in rotation. They improve soil structure, biological activity and crop yields, as well as increasing soil organic matter and nitrogen. This Chapter describes those aspects of soil biology, organic matter and soil nutrition central to the observed benefits of crop rotations.

SOIL BIOLOGY AND THE BIOLOGICAL MICRO-ENVIRONMENT

Organic matter, microscopic and macroscopic organisms (e.g., fungal hyphae and invertebrates), detritus from fungi and animals, and bacteria, and biological exudates, all assist in stabilizing soil structure. The role of each part of the biomass differs according to its size. Broadly, large aggregates greater than 250 m m diameter (macro-aggregates), are stabilized by their inherent physical structure, wetting and drying cycles, and organic matter. Micro-aggregates (< 250 m m) are stabilized by live or dead roots, fungi, invertebrates and microorganisms.

The populations of soil organisms of all sizes are linked functionally through their roles in the degradation of various forms of organic material. The latter includes live and dead plant material and other live or dead organisms. This shows that animals such as nematodes and some fungi feed directly on live plants while other fungi and bacteria feed predominantly on litter.

Earthworms and other large invertebrates create, and inhabit, burrows and pores, and are very mobile. The most notable of these are termites, which are divided into three groups according to the structure of their nests: those that build mounds (a) above ground, (b) on the soil surface, and (c) below ground. Small arthropods, microfauna and fungi live mostly in larger voids and in association with roots. Foster (1988) reviewed the location of the various types of soil-dwelling organisms and found that fungi, which constitute about 80% of the biomass in many soils, tend to be restricted to the rhizosphere of roots, to larger pores between aggregates and to the surface of aggregates. Bacteria, by contrast, are found on roots in the rhizosphere, in small colonies in the larger micropores, within aggregates and on and within cell debris.

Organic Matter

Both plants and animals provide inputs of organic matter to soils. Once within the soil organic residues can be distinguished on the basis of their chemical structure (e.g., old lignified humic substances that degrade slowly), by their source (plant or animal) or by location. The standing crop of litter in semi-arid grasslands is usually more than 3 t/ha and in temperate dry steppe may exceed 11 t/ha (e.g., Klemmedson 1989).

There has been much debate about the relative contents of organic matter in tropical and temperate soils. Within those wet-and-dry climates that have hot summers assisting rapid decomposition, there is no evidence of inherently lower levels of organic matter in the tropics than in comparable temperate regions (Juo and Payne 1993). Kowal and Kassam (1978) and Juo and Payne (1993) review the role of organic matter in tropical soils. Here, it is sufficient simply to state that organic matter has various interrelated effects on soil fertility. In particular it should be noted that both chemical and physical effects are of relatively great importance in the soils of the semi-arid tropics because these generally have low cation exchange capacity. The relative importance of litter (crop residue) and manure as inputs of organic matter, varies between cropping systems and spatially within a system. Here, most of the above-ground crop residue is fed to animals, but an equal amount of below-ground crop material enters the soil organic matter pool.

Where alley cropping and agroforestry are practised, values are more variable, but possible inputs could be very significant where the trees, from which the litter is taken, are grown away from the annual crops. If, say, two-thirds of the leaves from leguminous trees are harvested annually, litter values will be substantially higher and the material of better quality than the leaf and stem residue from an annual crop likely to be recycled in the field. Some qualifications, however, should be made. Tree root material is not available for decomposition in the crop field unless it is spatially overlapping (e.g. as an intercrop), in which case the trees will compete with the crop for soil nutrients, water, light and space.

The proportion of animal and human manure used on cropland is more variable. Some farmers have developed stable systems which strongly emphasize the use of animal manure on crops. For example, Norman *et al.* (1982) describe how farmers in northern Nigeria managed to apply 4 t/ha of manure to their heavily-cropped crop land though they had only 3 cows each. Many other farmers do not ensure adequate recycling, either because they are more concerned with livestock management or they do not know the importance of maintaining a 'zero nutrient budget' to replace nutrients removed by the crop. For example, Norman *et al.* (1982) also describe farmers with 10 cows each, who applied only 1.9 t manure/ha to their crops.

Within a cropping system, manuring practice varies with location. There is transference towards the centre of the system. On traditional farms, the area near the household or village is highly fertilized with human and animal manure while more distant fields receive little or no organic matter. Fussell (1992) describes such a traditional 'ring' farming system in semi-arid west Africa. Here, if houses are thatched, the village needs rebuilding or moving every 2 to 4 years. Moving takes advantage of the fertility gradient. Where the huts are not moved, the fertility gradient becomes steeper with time. Rather than trying to even out fertility by labour-intensive transport of manure, farmers vary the cropping of the fields. Continuous cropping of millet is sustainable close to the hut or village where there is plenty of human and animal manure but crop rotations are essential at the periphery.

Biomass production of leaves from multipurpose trees (Source: collated in Young 1989)

Country	Land use	Tree	t/ha/year
Malaysia	Plantation	*Acacia mangium*	3.06
Philippines	Plantation	*Albizia falcataria*	0.18
Costa Rica	Hedgerow intercropping	*Calliandra calothyrsus*	2.76
Philippines	Plantation	*Gmelina arborea*	0.14
Indonesia (Java)	Plantation	*L. leucocephala, A. falcataria, Dalbergia latifolia, Acacia auriculiformis*	3.00-5.00
		Cordia alliodora	2.69
	Plantation crop	*C. alliodora + cacao,*	6.46
Costa Rica	combination	*Erythrina poeppigiana,*	4.27
		E. poeppigiana + cacao	8.18
Nigeria	Hedgerow intercropping	*Cajanus cajan*	4.10
Nigeria	Hedgerow intercropping	*Gliricidia sepium*	2.30
Nigeria	Hedgerow intercropping	*L. leucocephala*	2.47
Nigeria	Hedgerow intercropping	*Tephrosia Candida*	3.07
India	Plantation	*L. leucocephala*	2.30

Closed systems, which rely on maintenance of fertility by recycling organic matter, are not sustainable. Nitrogen levels may be maintained or even enhanced by using legume crops or pastures in rotation but even then further inputs of nutrients are required to replace those removed by the crop. Fussell (1992) describes the inevitable need to use inorganic nutrients or organic nutrients from off-farm sources such as wastes from cities to complement farm organic matter in the African Sahel: "Although the importance of organic manure in increasing cereal yields is well recognized, it is unlikely that there is sufficient available, on a regional basis, to sustain yields without the use of additional chemical fertilizers. For example in the Western African savannah 10 t/ha/year of manure are required for sustainable cropping of millet or sorghum...while in northern Nigeria 2.5 t/ha/year are sufficient to maintain yields in most areas." This four-fold difference largely reflects the integration of livestock with cropping in Nigeria which results in relatively high levels of soil organic matter. Farmers elsewhere, for example in Ghana, give lower priority to crop fertility, which lowers soil organic matter levels and leads to greater nitrogen loss through water erosion when there is little litter cover on the ground. Irrespective of these differences, however "It is doubtful if either of these levels of application can be achieved, as.. estimates cattle and sheep manure outputs in the northern savannah zone of Nigeria are estimated to be only 1.4 and 0.25 t dry matter per head per annum respectively." Off-farm sources of manure, organic matter generally (e.g., industrial by-products) and soil ameliorants will become increasingly important. Some, such as sewage from cities, are being researched. Their use in developing countries seems likely to increase, particularly as labour for transporting these materials is cheap relative to the cost of other nutrients such as manufactured inorganic fertilizers. The treatment and efficacy of various off-farm sources of organic matter are outside the scope of this *Bulletin*. It suffices to point out that substantial amounts of material are available.

Annual production of organic wastes in the USA and their current use on land

Organic wastes	**Total production**		**Current use on land[1] (%)**
	Million metric tons	**Percentage of total**	
Anumal manures	159	21.8	90
Crop residues	391	53.7	68
Sewage sludge and septage	4	0.5	23
Food processing	3	0.4	(13)
Industrial organic	7	1.0	3
Logging and wood manufacturing	32	4.5	(5)
Municipal refuse	132	18.1	(1)
Total	728	100.0	-

[1]Values in parentheses are rough estimates because of insufficient data.

Decomposition constants, k, for tropical legumes. Values were calculated with exponential equation for decomposition using data reported in the literature (Source: Juo and Payne 1993)

Species	**Location**	**Mean annual rainfall**	**Mean annual temperature °C**	**k/year**
Gliricidia sepium	Ibadan, Nigeria	1250	23-31	8.48
Flemingia congesta	Ibadan, Nigeria	1250	23-31	3.66
Cassia siamea	Ibadan, Nigeria	1250	23-31	2.17
Lonchocarpus cyanescems	Ibadan, Nigeria	1250	23-31	8.87
Inga vera	El Verde, Puerto Rico	4000	22	1.65
Inga sp. And *Erythrina* (mixed)	Caracas, Venezuela	1200	20	3.01
Erythrina sp. (mixed with non-legumes)	Caracas, Venezuela	1200	20	3.81
Inga edulis	Yurimaguas, Peru	2200	26	0.91
Cajanus cajan	Yurimaguas, Peru	2200	26	1.45
Erythrina sp.	Yurimaguas, Peru	2200	26	3.72

The rate of breakdown of litter and other organic materials both determines and depends on the populations and activities of organisms in the soil. It also determines the extent to which minerals taken up by a crop are released from its organic residues and made available as inorganic ions to the subsequent crop. There are numerous mathematical models of organic matter degradation. Almost all assume that the rate of degradation decreases with time, as the more soluble, digestible and accessible material is selectively and progressively degraded by soil organisms.

This selectivity is often arbitrarily described by recognizing several 'pools' of organic matter and, with time, 'passing' material from labile to less-labile compartments. In practice its physical and chemical composition changes gradually with time, as does its physical binding with soil particles and its relation to the predators that feed on it.

The rate of decomposition of leaf litter depends on its environment, particularly on soil temperature and soil water. Both these affect the physical breakdown of the litter and determine the population and activity of soil animals and fungi that feed on it. Decomposition also varies with plant type and age of litter, being slower for heavily lignified material. The specific properties of litter from different species, and the generally exponential form of litter decay (the rate of decomposition slowing with time) lead to values that suggest half-lives of litter ranging from 1 to about 10 years. Because of the importance of temperature in determining decomposition, rates decline from tropical to temperate locations within wet-and-dry climates.

Plant species, water and temperature are not the only factors affecting the speed of litter decomposition and mineral cycling. Management plays a role. There are four ways for managing crop residues: (a) stubble mulch in which residues are left standing; (b) surface mulch, where above-ground residues are cut and left on the top of the soil after harvest; (c) incorporation by ploughing; and (d) cut-and-carry, in which surface residue is removed and (if not used for livestock or thatching, etc.) returned as a surface mulch about planting time for the subsequent crop; this is usually combined with ploughing of below-ground residues.

Two of the above treatments involve the fragmenting of root residues. All of them affect a range of soil properties. The degree of contact between the residue and the soil (and its organisms) and other unexplained factors affect the rate of decomposition.

Suggests that the rates of release of minerals from organic matter, particularly litter, are a significant element in nutrient cycling and crop productivity in semi-arid environments. Such climates have a rainy season in which rainfall exceeds evaporation, so there is a risk of mineral loss through runoff and mineral loss and acidification by leaching. During the dry season there is potential loss of nitrogen as ammonia. This suggests the need to manage litter so most decomposition takes place when the subsequent crop is growing rapidly and is able to take up the nutrients released. It is wise at the same time to maintain ground cover if possible. Manipulation of litter requires labour, which may already be in maximum demand prior to sowing.

SOIL ORGANISMS ASSOCIATED WITH THE CROP

Bacteria and Nitrogen

Cropping in dryland regions needs nitrogen to be economically successful (e.g., Keating *et al.* 1991). Two sources of nitrogen are from organic matter and from nitrogen-fixing bacteria associated with plant roots. *Bradyrhizobium* and *Rhizobium* species infect plant roots forming galls or nodules, and fix nitrogen from the soil atmosphere directly to the plants. Locally-adapted, heat-tolerant strains survive from crop to crop in wet-and-dry climates and, whether established by natural colonization or by inoculation of the crop seed at sowing, they subsequently fix variable quantities of nitrogen. Where bacterial infection is effective, the bacteria commonly fix between 70-100% of the nitrogen used by the crop, though the proportion is lower when the crop is given inorganic fertilizers.

Effect of soil mineral N and N fertilizers on crop N productivity and the proportion (P) and amount of crop N derived from N_2 fixation (Source: Peoples and Craswell 1992)

Species	Location	Level		Total crop	N_2 fixed	
		Soil mineral N (kg N/ha)	Fertilizer N (kg N/ha)	N (kg N/ha/crop)	Proportion	Amount (kg N/ha/ crop)
Groundnut	India	-	0	196	0.61	120
			100	210	0.47	99
			200	243	0.42	102
Chickpea	Australia	10 (to 120 cm)		114	0.85	97
		326		184	0.17	33
			0	109	0.80	87
			50	110	0.55	60
			100	104	0.29	30
Soybean	Australia	70 (to 120 cm)		230	0.34	78
		260		265	0.06	16
	India	-	0*	63	0.29	18
			100	148	0.26	28
		-	0**	89	0.48	43
			100	115	0.24	28
Common bean	Kenya	-	10	149	0.39	58
			100	158	0.10	16
Cowpea	Kenya	-	20	116	0.53	62
			100	137	0.08	11
	India	-	0	163	0.77	125
			100	138	0.67	92
			200	172	0.33	57

* Uninoculated.

** Inoculated.

The extent of the effectiveness of infection of legume crops in the wet-and-dry tropics needs to be surveyed. Temperate research indicates that nitrogen fixed by bacteria ranges from 20 to 120 kg N/ha in a growing season for annual crops. In the semi-arid tropics, amounts of nitrogen fixed per hectare range from none, where nodulation is ineffective, to 16 kg N in soybean naturally colonized by rhizobia, to 84 kg N when inoculated. Nitrogen fixation in soybean and groundnut of 50-70 kg N/ha/growing season is reported in Senegal (Gigou *et al.* 1985). Nitrogen-fixing bacteria associated with tree legumes can fix similar quantities.

Rhizobia rarely fix all the nitrogen used by the crop and a substantial amount of nitrogen is removed from the field as grain, so the net benefit from symbiosis varies.

For example, if rhizobia fix 50% of the crop's requirement and 60% of above-ground nitrogen is in the grain at harvest, soil nitrogen is depleted irrespective of the amount of nitrogen fixed or the grain yield.

Peoples and Craswell (1992), studying 21 cases of tropical grain legumes, found positive nitrogen balances in 15 and negative ones in five. Similarly, many experiments with grain legumes in southern, semi-arid Australia show that rhizobial associations give net benefits in more than three-quarters of cases and that the benefit from lupins is substantially greater than that from field peas, largely because a smaller proportion of nitrogen is harvested and removed in lupins.

FUNGI, ALGAE AND NUTRIENTS

Various fungi facilitate uptake of nutrients by plants, particularly phosphorus. Numerous fungi live in close association with plant roots. One group, vesicular arbuscular micorrhizal fungi (VAM), form both vesicles and arbuscules (knot-like structure) on the surface and within the root. They also colonize soil animals including earthworms and woodlice.

They are most prolific in the topsoil, to about 10 cm depth (Habte 1989). They facilitate uptake of nutrients, particularly phosphorus from soils low in that element.

They provide some protection to the host plant, their presence being associated with decreased colonization by pathogens. Ellis *et al.* (1985) also found that wheat plants inoculated with VAM were more drought-tolerant than plants without VAM. Importantly, comparisons of conventional cropping systems using inorganic fertilizers and herbicides with organic systems not using herbicides have found much higher levels of infection of crop roots by these beneficial fungi in the organic system (Ryan *et al.* 1994). This suggests that management for desired, associative soil organisms, should become a part of future sustainable cropping.

Nitrogen fixation by trees and shrubs. Values are per growing season or per year unless the number of months is given in brackets (Various sources, compiled by Young, 1989 and Peoples and Craswell 1992)

Species	**N fixation (kg N/ha/year)**
Acacia albida	20
Acacia mearnsii	200
Allocasuarina littoralis	220(?)
Casuarina equisetifolia	60-110
Coffee + *Inga* spp.	35
Coriaria arborea	190
Erythrina poeppigiana	60
Gliricidia septum	13
Inga jinicuil	35-40
Inga jinicuil	50
Inga jinicuil	35
Leucaena leucocephala	100-500
Leucaena leucocephala (in hedgerow intercropping)	75-120
Leucaena leucocephala	100-13 (6)
Prosopis glandulosa	25-30
Prosopis glandulosa	40-50
Prosopis tamarugo	200
Rain forest fallow	400-100
Mature rain forest	16
Alachynomene afrospera	130 (2)
Glyricidia sepium	100 (3)
Sesbania spp.	125-140
	119-188 (2)
	140-290 (2)
Xalliandra colothyssus	11 (3)

Other organisms associated with or infecting plants include endophytic fungi which affect plant growth rates and confer resistance to some insects. Cyanobacteria, which belong to the blue-green algae, are believed widespread in the semi-arid tropics and contribute 0.5 to 15 kg N/ha/year. *Azospirillum, Klebsiella, Enterobacter* and other species are free-living but largely inhabit the root rhizosphere; their distribution and role is still to be defined for cropping systems in wet-and-dry climates. Actinorrhizal associations are known between *Casuarina* and *Frankia* sp., but these are not exploited. It is claimed that such associations can fix as much as 60 kg N/ha/year. There are no doubt other symbioses yet to be discovered and exploited.

Weeds, Crop Pests and Diseases

Weeds, pests and diseases all compete with or directly reduce the vigour of crops. Many pests and diseases are soil-borne. Weed life-cycles depend on replenishment of the soil seed bank and survival of the seeds against natural decay, predation by soil animals and depletion by human management, particularly cultivations. Ecological weed control thus aims to minimize recruitment of new seed into the soil as a long-term strategy as well as trying to reduce artificially the size of the weed seed bank in the soil.

Frequency (%) of VAM fungi in soil-dwelling macro-invertebrates sampled from natural and agricultural ecosystems in Ohio (data are combined from 1986 and 1987 samplings) (Source: Rabatin and Stinner 1989)

Taxa	Ecosystem			
	Conventional tillage maize	No-tillage maize	Pasture	Old field
Lumbricidae (earthworms)	25.0	83.3	50.0	75.0
Isopoda (woodlice)	100.0	35.7	64.7	36.8
Carabidae (ground beetles)	2.1	19.8	14.5	12.8

Weed life cycles and strategic weed control are well understood for temperate wet-and-dry climates. The weed populations were measured under the contrasting conditions of slash-and-burn and terrace cultivation. Seed recruitment to the soil seed bank was higher under terrace cropping than under slash and burn but fewer survived.

It is generally accepted that there are links between weed populations, their competitiveness, and soil fertility. A good example of this is the various species of witch weed (*Striga hermonthica, S. generioides, S. asiatica* and approximately 25 other species). These species infect the roots of millet, sorghum, maize and sugar cane. *S. generioides* also infects legumes and tuberous crops.

They are a serious nuisance and are widespread in semi-arid Africa. It is estimated that they currently infect 21 million ha and a further 44 million ha are at risk. Their spread and increased aggressiveness is considered to be due largely to declining fertility and soil carbon contents.

The incidence of insect pests and crop diseases depends on the complexity of the cropping system and the types of crops grown. It depends too on management factors, such as the timing and type of tillage and the treatment of the crop residues. These aspects are so complex they warrant a Bulletin to themselves, so they are only touched upon here.

The incidence of pests and diseases is generally believed to be less frequent and/or intense in diverse cropping systems. Data on insects are relatively scarce but Andow (1983) is frequently quoted as showing a positive relationship between insect pest populations and the practice of monoculture. His experience, however, is mainly with cotton, a unique crop, and wheat, though he gives a few examples from maize and rice.

Viruses for which disease incidence is lower in more diverse cropping systems (Source: various, collated in Power 1990)

Pathogen	Type[1]	Vector
Alfalfa mosaic virus	N	Aphid
Bean common mosaic virus	N	Aphid
Bean yellow mosaic virus	N	Aphid
Beet yellows virus	P	Aphid
Cauliflower mosaic virus	N	Aphid
Chlorotic mottle virus	N	Aphid
Maize stunt spiroplasma	P	Leafhopper
Cucumber mosaic virus	N	Aphid
Groundnut rosette virus	P	Aphid
Pepper veinbanding mosaic virus	N	Aphid
Tomato yellow leaf curl virus	P	Whitefly
Turnip mosaic virus	N	Aphid

[1] Type of pathogen transmission system is given as P (persistent) or N (non-persistent).

Power (1990) shows that the incidence of virus diseases is less in complex cropping systems. He suggests that disease transmitted by insects is less common in multiple-cropping than in monocultures because aphids and some other insects settle on plants in response to visual cues that do not enable them to distinguish between host and non-host plants. In this way polyphagous insects can acquire a pathogen from an infected plant and transfer it when feeding on a non-host plant. If the insect is preferentially attracted to the non-host crop, it may be speculated that this crop acts as a 'trap' or 'protection' crop.

Power (1990) reviews situations where intercropping or the inclusion of a second, barrier species has led to lower disease levels. In contrast, bacteria and fungi in the root region and rhizosphere are reported to be slightly fewer in a coconut monocrop than in a multi-storied mixed cropping system (Bopaiah and Shetty 1991). Differences in populations, and in their varied (positive or negative) effects on the food crop, are no doubt best understood through analysis of the feeding habits and predation of the organisms themselves, rather than by making superficial generalizations about their numbers. Fick and Power (1992) and Hearn and Fitt (1992) review pest management generally and pests and diseases in cotton crops specifically.

INORGANIC NUTRITION

Inorganic nutrients occur in soil as ions and minerals, e.g., as oxides, silicates and phosphates, both adsorbed onto the surface of clay particles and organic matter, and in solution.

A large proportion of some nutrients, most notably nitrogen, is found in organic matter in all but the most highly degraded soils, so organic matter and the organisms associated with it are given prominence in this chapter.

Clay particles, because of their crystalline structures, carry an inherent electrical charge. This results in attractive forces (mainly van der Waals forces) and repulsive forces (electrostatic forces) which give clay species their particular characteristics. The inherent surface charge also causes a layer of associated ions to align next to the solid particles forming a so-called diffuse double layer because it consists of a relatively inexchangeable layer (the Stem layer) closest to the surface of the particle and an outer, readily exchangeable layer, of varying thickness, called the diffuse layer. Sposito (1984) explains this more fully.

In the present context, it is appropriate to describe briefly four aspects of the physico-electrical structure of soils. These are: the inherent exchange capacity; the linkage between soil particles and soil organic matter, and its effects on soil stability; ions in solution and their solubility; and the ion imbalances (from the perspective of crop growth) associated with acidity and salinity.

Representative cation exchange capacities (in mol_c/kg) of surface soils (Source: Sposito 1984)

Soil order	CEC
Alfisols	0.1 2 ± 0.08
Aridisols	0.16 ± 0.05
Entisols	0.13±0.06
Histosols	1.4±0.3
Inceptisols	0.19±0.17

Mollisols	0.22±0.10
Oxisols	0.05±0.03
Spodosols	0.11 ±0.05
Ultisols	0.06 ±0.06
Vertisols	0.37 ± 0.08

Adsorption is the net accumulation of matter at the interface between a solid phase and an aqueous solution. Readily exchangeable ions are those so loosely adsorbed to clay or other particles that they are replaced easily by leaching with an electrolyte solution.

The ion exchange capacity of soil is the number of moles of adsorbed ion charge which can be displaced from a unit mass of soil; in most applications, this refers to readily exchangeable ions. Sposito (1984) notes that 'Much controversy exists over the surface chemical significance of ion exchange capacities'.

The maximum surface charge measurement indicates a soil's potential to adsorb ions while its actual capacity, which is more relevant agriculturally, has a lesser value. It is notable that though the CEC of each order ranges widely, the predominant soils in wet-and-dry climates have low CECs.

The readily exchangeable proportion of the CEC varies with soil pH: the proportion available to readily-exchangeable bases decreases from about 1 at pH 8 to 0.5 at pH 6 to perhaps 0.2 at pH 4.5. Below pH 6 an increasing proportion is taken up by various aluminium ions and complexes, which are variably toxic to plants.

The electrical conductivity and CEC of a soil are related to its clay content (e.g., Rhoades 1990a). Similarly, because of the electrical charge of the clay particles, a high but variable percentage of the soil organic matter is bound to them.

That it may be as high as 90% and illustrates the two postulated main ways that organic matter is bonded to clay. These are weak anion exchange and strongly-held ligand exchange which is a form of chemical bonding.

It follows from the electrically-charged nature of clay and organic matter and the high likelihood of their association that both contribute to the nutrient-holding capacity and stability of soils. This intimacy led Pieri (1992) to propose that structural stability, at least of the African soils he studied, could be described by a critical value of S greater than 9, where S is the ratio of organic matter to clay plus silt, expressed as a percentage.

Proportion of soil organic carbon in the clay-organic complex

	Percent organic carbon in soil	Percent soil carbon in clay-organic complex[1]
Podzol	1.5	18
Solonized brown soil	1.5	77
Grey clay soil	1.1	91
Terra rossa	2.8	82
Groundwater rendzina	5.4	69
Black earth	1.8	82
Krasnozem	4.9	90
Red brown earth	2.5	66

[1] Defined as the material sinking when the soil was ultrasonically dispersed in an organic liquid of density 2 g/cm^3.

Plants take up most minerals as inorganic nutrients from the soil solution (except legumes, which may fix dinitrogen gas directly from the soil air). An element is available (to the plant) if it is present as, or can be transformed into, a free ion, and it is within the plant root zone. Most elements move within physical proximity of roots through soil water movement and into the plant through evapotranspiration.

Diffusion along concentration gradients is important for less-mobile ions such as phosphorus, particularly where soil solution concentrations are weak and root densities are high (i.e., the transport path is short). Sposito (1984) and others give calculated values for the diffusivity of nutrients. Diffusion times range from 1 day for an ion to move 3 mm (which is comparable with the time it would take to move by convection in the mass flow of water) for nitrate to about 200 days for potassium, magnesium and molybdenum, and to thousands of days for other nutrients.

The concentrations of nutrients in solution fluctuate daily and seasonally. The most dynamic are nitrate and ammonium ions, which are interconverted by bacteria. Fluctuations can be explained by the effects of temperature and soil water on mineralization (breaking down organic matter to release ammonium), immobilization (the reverse) and nitrification (conversion of organic matter to nitrate, which is stable and highly soluble), and by the effects of rainfall leaching nitrate to depth. In seasonally wet-and-dry cropping systems, there is a flush of nitrate following the start of the wet season and nitrate-N may accumulate in the topsoil by capillary rise of water during the dry season. Within the major seasonal patterns driven by soil water, Rochester *et al.* (1991) found that the soil nitrate cycle lagged three months behind changes in temperature.

Relative salt tolerance of various crops at emergence and during growth to maturity (Source: Rhoades 1990b)

Crop		Electrical conductivity of saturated-soil extract	
Common name	**Botanical name**	**50% yield (dS/m)**	**50% emergence (dS/m)**
Barley	*Hordeum vulgare*	18	16-24
Cotton	*Gossypium hirsutum*	17	15
Sugarbeet	*Beta vulgaris*	15	6-12
Sorghum	*Sorghum bicolor*	15	13
Safflower	*Carthamus tinctorius*	14	12
Wheat	*Triticum aestivum*	13	14-16
Beet, red	*Beta vulgaris*	9.6	13.8
Cowpea	*Vigna unguiculata*	9.1	16
Alfalfa	*Medicago sativa*	8.9	8-13
Tomato	*Lycopersicon lycopersicum*	7.6	7.6
Cabbage	*Brassica oleracea capitata*	7.0	13
Corn (maize)	*Zea mays*	5.9	21-24
Lettuce	*Lactuca sativa*	5.2	11
Onion	*Allium cepa*	4.3	5.6-7.5
Rice	*Oryza sativa*	3.6	18
Bean	*Phaseolus vulgaris*	3.6	8.0

Low soil pH affects plant roots directly because of the effects of the hydrogen ion concentration on root membrane integrity and exchange capacity. Acidity also affects roots indirectly in two ways. It alters the availability of ions in the soil solution (possibly making available toxic aluminium species, relatively toxic to plants). It also affects mineralization through the protons competing with cations for dissolved ligands and surface charged groups.

Soil pH also affects microorganisms, and thus the speed of transformations, for example, those between nitrate and ammonium. Poor plant growth on acid soils may thus be caused directly by hydrogen ions, by toxicities of aluminium or manganese, or through deficiencies of calcium, magnesium, potassium, phosphorus, nitrogen or trace elements. Chase *et al.* (1989) describe these relationships for sandy Sahelian soils. Variation in pH across distances of 15 m can be as much as pH 4.5 to 7.5, with associated decreases in aluminium and hydrogen ions, and increases in crop productivity. The critical pH at which crop growth is affected varies with crop, cultivar and soil type. Critical levels may be as high as pH 5 to 5.5 for less tolerant plants in soils with soluble sources of aluminium but otherwise can be as low as pH 3.9 to 4.

Saline soils and saline-sodic soils have electrical conductivities greater than 4 dS/m. The saturated soil extract of saline soils contains more than 15% exchangeable sodium; that of sodic soil less than 15%. The pH values of both kinds of soils range between 7 and 10.5 directly reflecting the amount of sodium bicarbonate present, this being completely dissolved in moist soil.

FIELD INDICATORS OF BIOLOGICAL AND NUTRITIONAL PROBLEMS

In this chapter, whilst reviewing relevant recent work on soil organic matter, soil biology and inorganic soil nutrients, a non-traditional perspective has been taken by placing emphasis on soil organic matter and the close chemical and functional relationship between organic matter and soil particles. It is therefore consistent to suggest that the most obvious indicator of biological and nutritional problems is the absence of, or reduction in, organic matter. Rates of loss of organic matter independent of erosion tend to be slow, but important as they are cumulative.

Larson and Pierce (1991) propose that minimum, analytically-gathered data sets are essential for monitoring soil sustainability. They include two measures of organic matter among the ten attributes that they consider essential. Once the minimum data set is constructed (Larson and Pierce 1991), each soil attribute is determined for a reference time (T0) and the change in soil conditions can then be measured over time (T1), which they suggest should be one to ten years.

While suggesting that agreed repeated observations or measurements are essential, it is preferable to empower farmers to take responsibility for recording their own observations of surrogate measurements rather than giving responsibility for analytically-based measurements to outsiders.

Policy-makers need to address the need for analytical or surrogate assessments of soil sustainability. A first step would be to ask farmers for their observations of soil characteristics and their attitudes and concerns about soil productivity.

In the case of soil organic matter measurements, it is believed that the limited availability of expertise, the costs and the inherent sampling errors make direct measurement of organic carbon neither cost-effective nor informative. It is more likely that surrogate measures are adequate and, if sufficiently simple and cheap, have some likelihood of being used. Pieri (1992) suggests that a bleached (possibly brittle) soil surface and plant deficiency symptoms are useful surrogates for loss of organic matter, low CEC and soil nutrient imbalances.

Indicators of plant nutrient deficiencies include: colour (e.g. redness for potassium deficiency); paleness (a more general symptom, but especially obvious in nitrogen deficiency); and small leaves and plants. Other symptoms, such as leaf curling and accelerated dropping of older leaves, may also be helpful but might equally indicate water deficits. There are several publications with photographs of nutrient deficiencies; these, however, might require further tailoring to specific combinations of crops and soils.

Soil acidity is most obviously indicated by stunted plants, by barc ground, and by better relative growth of acid-tolerant weeds in a sensitive crop. Salinity too is indicated by vegetation changes such as trees dying for no apparent reason or an increase in the proportion of salt-tolerant herbs. Other symptoms of salinity include: waterlogged or bare soil; livestock congregating and licking the soil surface for salt; visible salt crystals; the smell of salt; and clear catchment water because salt settles sediment.

INTERPRETATION OF ELECTRICAL CONDUCTIVITY OF SOILS

In actuality, the interpretation of EC of a soil or media must be made considering the plant(s) to be grown. The EC of the soil has little direct detrimental effect on sandy mineral soils or on media. However, EC directly affects plants growing in the soil or media. The impact of EC on plants is also directly affected by water management.

Salt Index Use and Calculation

As EC increases, more attention to water management is needed to prevent salinity from adversely affecting plants. The Extension Soil Testing Laboratory uses a 2:1 solution:soil ratio with which to determine EC. Many states use a saturated paste extract. This saturated paste method is more time consuming than the 2:1 extraction, and results in inadequate amounts of solution in Florida's sandy soils. The conversion from the 2:1 extraction result to the saturated paste result, termed salt index, is easy and accurate.

EC (salt index) = EC (2:1) x 8.

In general, when the soil EC (2:1, water:soil) exceeds 0.25 dS/m (or 0.25 x 8 = 2.0 dS/m salt index), many plants experience stress due to salts. Other plants (e.g., bermudagrass) are quite tolerant to salts. Due to this species-dependent effect of EC, a listing of the effects of increasing EC on selected plants has been compiled (Hanlon et al., 1993).

EC Probe and Meter

There are a wide variety of EC probes and meters marketed. Those instruments using probes that are placed in the soil or media *in situ* (directly in the soil without taking a sample) are not considered in the following discussion. Since EC is a measurement of the conductivity of the soil solution, the measurement should be made under a controlled mixture of solution to soil (2:1). These conditions usually do not exist when direct reading instruments are used.

Electrical Conductivity Probe

The probe consists of a tube, usually of plastic, into which electrodes have been installed. Two common electrode arrangements are: 1) two plates; or 2) a rod located concentrically in a ring. In either case, the electrodes are held a specific distance apart. The gap between the electrodes is filled with the water sample or filtered solution from a soil sample, either by filling a reservoir or by placing the probe in the solution. When the probe is immersed in the solution, ions contained in the solution will permit electrical flow from one electrode to another. If a large number of ions are present (salty conditions), then the EC of the sample will be higher than a sample with low number of ions (low salts).

Some older instruments require that a mixture of soil and solution be packed into the electrode receptacle. While readings from these older instruments may be adequate, accurate results are harder to obtain because readings are affected by the sample packing method.

Inspection of the probe should include insuring that the probe is clean and free of debris, and that all electrical connections are in place. Some probes may experience a buildup of corrosion on the electrodes with time. Indication of corrosion, if not directly visible, is indicated by the constant need to reset the instrument because of drifting, often in one direction as corrosion progresses.

Electrical Conductivity Meter

Meters for EC are extremely reliable. Meters from the 1950s which use a "cat's eye" tube are still in use. Newer meters using digital displays are often susceptible to failure of one or more segments of the digital display, usually related to corrosion within the instrument.

As with failures associated with pH meters, EC meters will always provide an EC, unless the problem results in total instrument failure. It falls upon the operator to insure that the reported EC is an accurate measurement of the conductivity of the sample.

LABORATORY PROCEDURES

The following procedures are used at the ESTL and are based on good laboratory procedures with sufficient quality control measures to ensure that pH and EC readings are accurate and reliable. While equipment may vary among County Extension Laboratories, procedures should be developed that directly parallel those used by the ESTL.

Soil Scooping Technique

The ESTL uses a scoop (that is, a volume\ measurement) for both pH and EC determinations. The scoop is a plastic and metal device which may be obtained from a commercial manufacturer (e.g., Custom Laboratories, Orange City, FL). Alternately, the scoop may be constructed from locally available materials, such as measuring spoons or coffee scoops. The intent is to use a consistent volume of soil and water.

The soil scooping technique requires practice, despite its unsophisticated appearance. The technique depends upon uniform actions by the technician from sample to sample to produce consistent packing of soil into the scoop. This consistency can be directly measured by repeatedly scooping the same soil and weighing each scoop. Weights should be uniform within each scoop (volume) and soil-sample combination. Weights will vary from soil to soil, especially when there is a noticeable difference in soil texture.

Sample Handling and Preparation

The sample should be air-dried and passed through a 2-mm sieve before scooping.

Procedure

1. Dip the scoop into the center of the soil sample and fill the scoop with a twisting motion so that extra soil is mounded above the rim of the scoop. Do not press the scoop or force the soil against the side of the container (Jones, 1980).
2. Strike the handle near the scoop three times with a plastic rod to settle soil particles.
3. Level the scoop with the plastic rod. Strike off all excess soil above the

rim of the scoop in a single stroke so that the soil is not compacted into the scoop.

Estimation of Soil Texture

Knowledge of soil texture is useful in the recommendation of lime. Local soil conditions may be quite uniform so that little differences may be found throughout the county. However, such uniformity is expected to be rare in most counties. After the soil has been air-dried and sieved, a small amount of dry soil should be moistened and rubbed between the forefinger and thumb. An estimation of the texture is made by comparing the"feel" of the sample to that of a set of soil samples of known texture.

Soil pH (2:1 V/V)

This procedure uses a 20-cc (~25-g) soil scoop and 40 mL of pure water to obtain a 2:1 water-to-soil ratio. Most problems with this procedure are associated with the glass or calomel electrodes.

However, sample pH may also be affected by contaminated water, by microbial activity if samples are allowed to sit for several hours before determining pH, or by improper scooping techniques.

1. *Standard Solutions:* Obtain commercial standard solutions of pH 4.00, 7.00, and 10.00.

   Sample Handling and Preparation: The soil sample should be air-dried and passed through a 2-mm sieve. Irrigation water samples require no preparation.

   Procedure: 1. One scoop of soil to a 3-oz plastic cup using a 20-cc (~25-g) scoop.
2. Add 40 mL of pure water to each cup using an automatic pipette or suitable volumetric container. Stir with a glass rod and let the sample stand for 30 min.
3. Standardize the pH meter.
4. Stir the sample again immediately before measuring the soil pH. Do not place the electrode(s) directly in the sand layer at the bottom of the container. The electrodes should be positioned in the solution just above the sand layer.
5. Record pH to the nearest 0.1 pH unit (suggested format of XX.X).

Electrical Conductivity

This test, often called "Soluble Salts," requires that 20 cc (~25 g) of soil be mixed with 40 mL of pure water, resulting in a water:soil ratio of 2:1. The

4-hr equilibration period provides time for some slowly-soluble constituents to approach solution equilibrium. Little error results from much longer equilibration times, but shorter time periods might introduce inconsistent results for some samples.

Standards

A solution of 0.005 N KCl has an electrical conductivity of 720 ± 1 dS/m (mmho/cm) at 25°C. Use a commercially prepared solution as the reference solution.

Sample Handling and Preparation

- The soil sample should be air-dried and passed through a 2-mm sieve. Irrigation water samples require no preparation.

Procedure

1. Place 20 cc (~25 g) of soil in a plastic 3-oz cup.
2. Add 40 mL of pure water, stir, and allow to stand for 4 h.
3. Without stirring the sample, filter the solution through a Whatman No. 41 (11 cm) paper and collect the extract in a funnel tube or other suitable container. The intent is to remove the soil and other debris from the solution.
4. Standardize the conductivity meter.
5. Move the probe up and down in the solution several times to dislodge any bubbles on the electrode surfaces. Measure the electrical conductivity of the extract contained in the funnel tube.
6. Rinse the interior and exterior of the probe with pure water between samples. Remove any excess water from the exterior of the probe by blotting with a tissue.
7. Record all meter readings as displayed. Note that some meters use a floating-point display while others use a reading which depends upon a switch setting.

pH Meter

Instructions for the proper use of a pH meter, including electrode care and instrument-specific settings, are given in this subsection. While instrument settings may vary among meters, daily operation of pH meters is relatively standard. Most problems with pH instruments originate within the electrode(s) or in the electrical connections to the instrument. Prevention of early electrode deterioration is best accomplished by following the information in sections

B and C that follow. Use of only one standard buffer solution is not adequate for proper calibration. Do not reuse a buffer solution which has been left out overnight or has been used to "store" electrode(s). It is advisable to maintain a supply of a reference soil sample to be used as a check sample with every set of samples. Early detection of problems is the best method of avoiding inaccurate results.

Calibration of Electrode(s)

1. Rinse electrode(s) with pure water and blot dry with a tissue or clean paper towel.
2. Using commercially prepared buffer solutions of pH 4.00, 7.00, and 10.0, pour approximately 30 mL into labeled, 3-oz. plastic cups. Do not reuse buffer solutions in which electrode(s) have been immersed for daily storage.
3. Set the meter to operate in the "pH" mode.
4. Place the electrode(s) into new buffer solution (pH 7.00) and allow the electrode(s) to equilibrate. A stable reading should be obtained within about 30 seconds. Readings which are not stable after 1 minute indicate that the electrode(s) may be malfunctioning.
5. Using the Calibration" (often labelled "Standard") knob, adjust the pH-meter display to read 7.00. Do not move (or press) either the "Temperaturer" or "Slope" adjustments at this time.
6. Place the meter in the "Standby" mode. Some meters do not have a standby mode.
7. Rinse the electrode(s) into a waste cup with pure water and blot dry as above.
8. Place the electrode(s) in new buffer solution (pH 4.00) and allow the electrode(s) to equilibrate with the buffer solution.
9. Set the meter to operate in the "pH" mode.
10. Using the "Temperature" knob (sometimes labelled "Slope"), adjust the pH-meter display to read 4.00. Do not readjust the "Calibration" setting at this time.
11. Repeat Steps 6 through 9 using the pH 10.0 buffer solution. No settings should be changed as the pH 10.0 buffer is being read.
12. Repeat Steps 3 through 11 until readings of 7.00, 4.00, and 10.0 are obtained without adjusting the instrument. These standards should be set ±0.05.
13. Using a reference soil sample, read the pH and determine if the pH reading agrees with the "known" reading of the reference soil. Use of

a reference soil sample, a soil that can be analyzed with every sample set, is strongly recommended. Use of a reference soil is an excellent quality assurance measure and verifies that the electrode(s)/meter are functioning correctly in a soil solution and not just in standard buffers. The reference soil sample should be read once about 20 samples. If the reading drifts by more than ±0.2, then restandardize the meter.

14. Standardize the instrument according to the above procedure after every 50 samples. If the instrument has shown drift of 0.05 pH units or more, reread the last few samples to verify the accuracy of the recorded readings.

Daily Electrode Storage

1. Place the meter in the "Standby" mode.
2. Wash the electrode(s) with pure water.
3. Immerse the electrode(s) in pH 7.00 buffer solution.

Electrode Servicing

1. If the pH reading drifts, replace the glass electrode. Wait until the reading from the new electrode is stable. Usually, new electrode(s) should be kept in pH 7.00 buffer solution overnight before use.
2. After use, wash the electrode(s) with pure water.
3. Immerse the electrode(s) in pH 7.00 buffer solution.

Electrical Conductivity Meter

Conductivity measurements must be made on solution samples only. The most common problem with conductivity meters concerns failure of the electrodes (mounted within the hollow plastic probe) to make proper contact with the solution. The probe must be kept clean by adequately flushing the interior of the probe with pure water. When analyzing a solution, insure that the solution covers the bottom 3 to 5 cm (1 to 1.5 inches) of the probe. Agitation of the probe in the unknown and standard solutions is required to obtain reproducible results. Agitation insures complete wetting of the internal electrodes by removing any air bubbles using the solution of interest.

Calibration

1. Turn ON the "Supply" or "Power" switch.
2. Turn OFF the "Temperature Correction" switch (back panel).
3. Set the "Scale" wafer switch, if any, to read mmho/cm.
4. Place the conductivity probe into a standard solution of 0.005 M KCl.

Agitate the probe using an up-and-down movement to create better solution to probe contact.

5. After agitation, the meter should read 0.72 ± 0.04 mmho/cm (0.72 ± 0.04 dS/m).
6. Agitate the probe again and reread the standard solution. Both the first and second readings should be the same value if the probe is in good contact with the solution.
7. Wash the probe with pure water on both inner (inject wash water through the hole at the top of the probe) and outer surfaces. Blot the probe dry. Do NOT rub the outer surface of the probe with the tissue paper.
8. Read the reference soil sample and verify that the current reading agrees with the known value for the reference soil.
9. Record the entire meter reading including location of the decimal point.
10. Return the "Scale" wafer switch to the original position before reading the next sample.

SOIL ELECTRICAL CONDUCTIVITY VARIABILITY

Soil electrical conductivity (EC) is a property of soil that is determined by standardized measures of soil conductance (resistance^{-1}) by the distance and cross sectional area through which a current travels. Traditionally, soil paste EC has been used to assess soil salinity (Rhoades et al., 1989), but now commercial devices are available to rapidly and economically measure and map bulk soil EC across agricultural fields. The Veris® 3100 (Veris Technologies, Salina, Kansas) measures EC with a system of coulters that are in direct contact with the soil. The EM38 (Geonics, Limited, Mississauga, Ontario, Canada) induces a current into the soil with one coil and determines conductivity by measuring the resulting secondary current with another coil. Both sensors have been demonstrated to give similar results (Suddeth et al., 1999)

The movement of electrons through bulk soil is complex. Electrons may travel through soil water in macropores, along the surfaces of soil minerals (i.e. exchangeable ions), and through alternating layers of particles and solution (Rhoades et al., 1989). Therefore, multiple factors contribute to soil EC variability, including factors that affect the amount and connectivity of soil water (e.g. bulk density, structure, water potential, precipitation, timing of measurement), soil aggregation (e.g. cementing agents such as clay and organic matter, soil structure), electrolytes in soil water (e.g. salinity, exchangeable ions, soil water content, soil temperature), and the conductivity of the mineral phase (e.g. types and quantity of minerals, degree of isomorphic substitution, exchangeable ions). Despite the multiple causes of EC variability, bulk soil EC measurements have been related to individual factors that limit soil use and productivity such as salinity (De Jong et al., 1979; Rhoades and Corwin, 1981), clay content at a depth of 15-m in New Wales, Australia (r^2 = 0.78; Williams and Hoey, 1987), depth of sand deposition along the Missouri River (r^2 = 0.73-0.94; Kitchen et al, 1996), depth to claypan in Missouri (r^2 = 0.73; Doolittle et al., 1994), and soil moisture content (r^2 = 0.96; Kachanoski et al., 1988). If soil EC maps have utility in production agriculture, 1) EC must be spatially structured, 2) spatial patterns must have temporal stability, and 3) EC must be related to factors of agronomic importance. The objective of this research was to determine the nature and causes of EC variability for several fields in Kentucky. Geostatistical analyses were conducted to examine the spatial and temporal variability of EC variability. Transect studies were conducted to determine the causes of EC variability.

7

Biologically Degraded Soil

It is practically impossible for man to produce stable humus synthetically. Man can only properly cultivate the field, supply organic matter and encourage the development of stable humus in the soil. Soil with stable humus must always be protected to maintain the fertility and productivity. Stable humus is the crucial center, the focus of the soil life cycle. Adhering to farming practices that assure the production of stable humus becomes the farmer's main objective. In this paper, the authors discuss the various steps involved in assessing the biological state of the soil, by calculating humus content.

INTRODUCTION

For nearly one hundred years soil science in most schools of higher learning especially in agricultural sciences has been primarily concerned with the physical and mechanical aspects of soil structure. Biological thinking or biologically degraded soil has become a major concern only in last few years. The new approach consider not only the physical properties and mineral structure of the soil, but also the process by which organic matter is transformed into humus by microorganisms. Good soil consists of 90% mineral and 10% bio-organic substances. The bio-organic parts consists of 70-90% humus, 10-30% roots, and active fraction, which constitutes living organisms and other, constitutes.

Some of the nutrients get lost through leaching due to continually wet weather, melting snow, flooding or through denitrification. Also each cultivated plant takes away nutrients from the soil, as soon as the crops are harvested. The substantial task for the farmer is to take care in returning nutrients taken from the soil through harvesting. The conventional farmer is uses water soluble, mostly salty chemical fertilizer. In contrast, the bio-organic farmer uses organic matter in the form of crop residues, and other wastes, and/or compost, in order to take of soil life, its proliferation and simulation to highest effectively.

Microbial soil compound plays one of the most important roles in the decomposition of organic litter. Soil organic matter includes plant and animal remains in various stages of decomposition, cells and tissues of soil organisms, and substances produced by the soil microbes. Well- decomposed organic matter forms humus-a dark brown, porous, spongy material with a pleasant, earthy smell. The occurrence of microorganisms in the soil depends on many factors e.g. on soil acidity, air and soil humidity, air and soil temperature, soil water, all meteorological factors etc. Seeing an importance of soil biology this study only considers the mineralization of organic matter without erosion. A distinction is made between non-resistant organic matter (i.e. fresh organic residues) and resistant organic matter (humus). The biological degradation dealt in this study is as mineralization of resistant organic matter mainly based and soil parameter, which calculates the decay rate of humus.

STUDY AREA

The chosen study are in this study is India, lies to the north of equator, between 8o4' and 37o6' North and 68o7' and 97o25'. East It is bounded in the south by the Indian Ocean, in the west by the Arabian Sea, in the east by Bay of Bengal, in the north-east, north and a part of the north-west by Himalayan ranges, and the rest of the north-west by the Great Indian Desert. Physiographically, India can be divided into seven regions: the Northern Mountains, constituting the Himalayan ranges; the great Indo-Ganagetic Plain; the Central Highlands; the peninsular Plateau in the states of maharashtra, Andhara pradesh, Karnataka and Tamil nadu; the East Coast; the West Coast; and the Island.

The average elevation of the plateau that dominates most of Peninsular India ranges between 300 and 1800 m. The river systems of India may be broadly classified into two groups: the Himalayana and the Deccan.

The main rivers of the Himalayan group, which are snow-fed, are the Indus, Ganga and Brahmaputra including their major of the country. The Deccan system include the west-flowing Narmada, Tapi and Periyar and the east-flowing mahandi, Godavari, Krishna, Penneru and kaveri.

The Republic of India's mainland stretches 3214 kms from north to south and 2933 kms from west to east. The total land area is 3.29 million sq. kms. Size wise it is the sixth largest country in the world. India exhibits a variety of landscapes and climatic conditions in the evolution of different soils and vegetation. It shows a string relationship among the soils, climate and vegetation.

MANAGEMENT FOR MAINTENANCE OF SOIL BIOLOGY AND NUTRITION

The aim of management should be to create balanced organic matter and mineral budgets. It should ensure that, over several years (a complete crop rotation), soil organic matter is not depleted and that nutrients added equal or exceed those removed by cropping or lost in various ways.

When managing organic matter farmers should recognize that the effects of animal and human manure, sewage sludge and plant residues last longer than those of green manure crops. Green manures, though valuable, usually last only one or two seasons because they are incorporated before they are mature and lignified.

Juo and Payne (1993) advise: 'In spite of the many proven benefits of soil organic matter, its management and recycling in an intensified, modem agro-ecosystem must necessarily revolve around two fundamental characteristics of the system, namely, the availability of organic material at the farm level, and the economic incentive for conserving and recycling organic matter'. To this may be added consideration of the benefits and costs of treating and transporting human wastes from cities. On-site organic matter, organic material brought in from outside and industrially-produced fertilizers will be used according to their benefit cost ratios and the attitudes of governments and crop managers.

Tillage reduces the frequency of VAM, at least in invertebrates. Tillage reduces earthworm populations 10-30-fold, both killing them directly and destroying their burrows (Brust *et al.* 1986). Haines and Uren (1990) show that differences in earthworm populations between tillage treatments and stubble burning are reflected by 50% differences in numbers of soil pores.

Tillage also effects the location, numbers and activity of microorganisms within the soil, depending on the type of tillage. For example a mouldboard plough that inverts topsoil has more effect than minimal tillage. Doran (1987) found microbial biomass and potentially-mineralizable nitrogen to be 54% and 37% higher respectively in the top 7.5 cm of no-till than ploughed soil.

These would be critical differences if translated to the semi-arid tropics (Keating *et al.* 1991). Microbial and fungal biomasses may be 20-70% higher after 1 and 2 years no-till (Doran 1980). However, conventional tillage, by burying topsoil, may cause microbial populations and their activities to be higher than under no-till at depths of say 7.5 to 15 cm (Doran 1980). Differences in soil organisms (e.g., microbial biomass) under differing tillage treatments are seasonal, generally being greatest in the period when the weather is most favourable for soil organism activity and plant growth (Carter and Mele 1992). Rasmussen and Collins (1991) summarize many data on the impact of tillage on soil properties and conclude that stubble-mulching with zero tillage conserves up to 2% more organic matter per year than ploughing.

Micro-arthropod populations in top 10 cm of clay soil 1 and 2 years after adding organic fertilizers in Italy. Additions were made to achieve 4% organic matter and inorganic N and P were to make all treatments equivalent to the manure (Source: Fratello* et al. *1989)

Taxon	Control	Fowl manure	Sludge	Green manure	Straw
One year after application					
Collembola	13 209	31 635	13 653	17 205	24642
Entomobryidae	2 109	4 329	2 109	222	2 331
Onychiuridae	1 221	333	2 109	222	1 221
Isotomidae	9 990	26 529	9 324	16 650	20 313
C. coecus	8 214	24753	7 659	10 101	18 204
C. thermophilus	0	333	888	0	0
I. Minor	111	111	0	4 995	0
F. Parvulus	0	111	0	0	0
Sminthuridae	0	222	0	111	222
Neelidae	0	111	111	0	555
Acarina	65 046	42 513	35 853	43 068	70263
Pauropoda	3 108	1 221	3 108	2 331	7 104
Microarthropods	83 361	80475	54834	65 157	103 674
Two years after application					
Collembola	4773	7 104	8 769	5 328	5 550
Entomobryidae	0	999	444	222	111
Onychiuridae	444	444	111	222	0
Isotomidae	2 886	1 998	6 771	4 107	4 551
C. coecus	1 887	999	2 442	3 108	3 885
C. thermophilus	0	222	3 996	222	111
I. Minor	222	222	0	333	0
F. Parvulus	333	222	0	0	333
Sminthuridae	1 221	3 552	1 221	777	888
Neelidae	222	0	0	0	0
Acarina	12432	28083	24753	27 639	13 764
Pauropoda	999	111	1 554	999	777
Microarthropods	19425	35 964	36 963	35 187	21 090

Cropping patterns and rotations affect soil nutrition. Diversity of cropping increases the number and variety of soil organisms and reduces pests and diseases.

Soil acidification and salination are extreme cases of nutrient imbalance and, unlike other deficiencies, cannot be corrected simply by adding mineral fertilizers. The management techniques to reduce acidification are: (a) to reduce the net proton production of the system by minimizing nitrate leaching (a special problem in seasonally wet-and-dry climates); (b) to avoid using ammonium fertilizers and reduce the accumulation of organic matter; (c) to lime the soil. Liming is a high cost option while adoption of perennial species (to avoid seasonal accumulation of organic matter and nitrate leaching) is less costly. By contrast, salinity is usually managed by strategic leaching. This is done with the sparing use of low-salt water when both soil moisture and the water table are low.

SOIL BULK DENSITY

Soil Bulk Density Determination

Soil weight is referred to as soil bulk density. Density is the mass of material contained within a given volume. This is the idea that a given size of box may be heavy or light, depending upon what kind of material it contains. If the box were filled with wood, it would be light when compared to having it filled with lead. The weight of water is the reference for density measurements: 1 gram of water=1 cubic centimeter (cc), & 1 cc water=1 ml. & 1 cubic foot of water=62.4 lbs.

The bulk density of a soil is the mass of dry soil per unit of "bulk" (total volume of soil or soil particles & pore space).

B.D. = mass of oven dry soil (grams) ÷ total volume of soil (cm^3)=grams/ cc

The bulk density takes into account the total soil volume (the space occupied by the solid particles plus the space occupied by the air of the pores or pore space). To determine the mass of the soil—since air does not have any significant weight—we can just weigh the oven dry soil on a balance. The volume of the soil can be determined by pouring the soil into a graduated cylinder and measuring the volume that it occupies.

The problem with this procedure is that structural aggregates may be crushed or compacted once they are removed from the soil and placed into the cylinder. A better procedure is one in which the aggregates could be removed from the soil and frozen exactly the way they were in the soil. Plastic fixatives enable us to do this. A large aggregate has been removed from the soil and is being fixed by dipping it into a saran solution. This will make the clod impervious to water. The weight of our clump of soil can be obtained by hanging it on the balance or placing it on the metal weighing tray.

The volume of the clod also needs to be determined. This can be obtained by weighing the sample again, only this time the sample will be in water. The weight of the sample now will be less, since the sample will be buoyed up by the amount of water it is displacing.

Archimedes most famous theorem gives the weight of a body immersed in a liquid, called Archimedes' principal. If you want to know more about the mathematician who determined that this procedure would work, go to Archimedes.

Thus, by subtracting the weight of the clod in water from its weight in air, we obtain the weight of the water displaced by the clod, which equals the volume of water displaced by the clod (because 1 cm^3 of water=1 gm. of water) and is equal to the volume of the soil clod.

Clod Method of BD Determination =Mass of Clod ÷ Volume of water displaced

Volume of water displaced=weight of water or the (gms clod in air)-(gms clod in water)=(grams or volume of water displaced)

Cylinder method of BD Determination=Mass of oven dry soil (gms) ÷ total volume of soil (cm^3)

Another method of determining bulk density is by using soil cores. These are obtained with a probe that forces a cylinder into the soil. The soil can be removed from this cylinder and cut into sections and placed in another cylinder for transportation back to the laboratory. Care needs to be taken to avoid compacting the soil during and after obtaining the core.

BD of soil (using core method)={(mass of soil + mass of core)-(mass of core} ÷ (volume of core)

volume of core=Pi (3.1416) $r^2 \times h$

r=radius of core; h=the height of core.

For the cores in the lab, r=2.45cm (½ of Diameter); h=5 cm

$V=3.1416 \times 2.45^2 \times 5=94.3$ cm^3;

Bulk density is a fairly easy determination that will yield information about the soil that will be significant for determining the soil's potential for plant growth or a building foundation. Remember it is the oven dry mass/volume soil. or g/cc.

SIGNIFICANCE OF BULK DENSITY

The bulk density of the soil will play an important role in determining if the soil has the physical characteristics necessary for plant growth, building foundations or other uses.

From the laboratory investigations you will obtain very high bulk densities for the Bt and sandy soil. The weight of the soil is important if you are going to be lifting it or hauling it long distances. One of the main reason for sod farms to be on organic soils is to reduce the cost of transportation because the organic soil is so much lighter than mineral soils. The organic sod is easier to handle. Most sod farms in Minnesota are on the peat soils north of the Twin Cities.

Sometimes we are interested in the weight of an acre of soil for erosion comparison purposes. Erosion of 5 tons per acre sounds like a lot. However the weight of an acre furrow slice on average is 2,000,000 lbs., so 5 tons per acre seems like a small amount. Five tons per acre is only about 34 thousandths of an inch thick (.034 inches). However, many areas have lost 10 to 100 times that amount. (Note: an acre furrow slice, or AFS, is a three-dimensional volume of soil which is one acre in area and 7 inches deep.)

Weight per AFS=BD × Volume

Example: Volume of soil=Soil depth × 43560 $ft.^2$ (area of 1 acre)

For conversion of g/cc to lbs/ft^3, multiply BD × 62.4 lbs/ft^3 (weight of ft^3 water)

How many tons are lost if a soil with a B.D. of 1.2 g./ cm^3 erodes 5 inches per acre?

answer: (1.2 × 62.4)=(weight in lbs/ft^3) X Volume: (where Volume=(5 in. ÷12 in/ft) × 43560 $ft)^2$ or (74.88 lbs/ft^3) × 18,150 ft^3=1.3 million lbs or divide by 2000 lbs / ton or weight=679 tons.

Soil Factors Impacting Tillage

Another factor similar to carrying soil is its potential to be moved short distances, such as in plowingor rototilling. The soil that is heavier will be more difficult to move. However, another factor is the ability of the soil to stick together, or the soil's consistence.

Clay soil is noted for its stickiness and large energy requirements in tillage. Farmers refer to it as "heavy," but they really mean it is difficult to plow, not that it has a high bulk density; clay soils generally will have a lower bulk density than sandy soils (clay=1.3 g/cc vs sand=1.6g/cc). Sandy soils have a higher bulk density, but are easier to plow since they have weaker consistence. Thus they are often referred to as "light soils." The lower B.D. of clays is due to their better aggregation.

Rototilling the soil reduces the bulk density by "fluffing" the soil. For a look at tillage implements go to Tillage Implements

Soil Consistence

Soil consistence is the soil's ability to cohere or stick together. The soil's consistence may be evaluated at three moisture conditions: air dry, moist, and wet. Moist consistence is evaluated by placing the soil between the thumb and forefinger and gently applying pressure. The ease with which a ped can be crushed determines the consistency.

Terms commonly used to describe moist consistence are:

Loose-Non-coherent when dry or moist; does not hold together in a mass.

Friable-When moist, crushes easily under gentle pressure between thumb and forefinger and can be pressed together into a lump.

Firm-When moist, crushes under moderate pressure between thumb and forefinger, but resistance is distinctly noticeable.

Plastic-When wet, readily deformed by moderate pressure but can be pressed into a lump; will form a "wire" when rolled between thumb and forefinger.

Sticky-When wet, adheres to other material and tends to stretch somewhat and pull apart rather than to pull free from other material.

Hard-When dry, moderately resistant to pressure; can be broken with difficulty between thumb and forefinger.

Soft-When dry, breaks into powder or individual grains under very slight pressure.

Cemented-Hard; little affected by moistening.

In air dry conditions, the resistance to rupturing when rubbed is measured. At intermediate moisture content, the soil's resistance to shearing forces by thumb and finger is noted. In the wet condition, its plasticity—ability to be molded and stickiness—are measured.

One significance of the B.D. of a soil is the amount of surface area a soil has. The more surface area, the more ability to retain water and nutrients. Notice that the two soils pictured below weigh the same, but are significantly different in surface area. Which soil has the higher B.D.?

COMPACTION, POROSITY AND SOIL TEMPERATURE

Soil Compaction

Soil compaction works something like the soil particles seen below. When the soil is randomly fluffed, as on the left, the amount of pore space will be 45 to 55%. When we push the soil particles closer together, we remove the pore space, and the bulk density will increase. Compaction increases the bulk density by reducing the pore space.

Compaction can also result in the changing of the proportion of pore sizes. Notice in the compaction diagram that compacted soils not only have a lower total pore space (resulting in higher bulk density), but also less macro pores and more micro pores. This will often result in excess water retention.

When we walk on the soil we compact it. Eventually the compaction prevents the growth of plants. This trail TRAIL behind the Soils Building shows the effect of compaction and samples from the soil will be used in the lab to determine the amount of compaction.

Compaction changes the amount and size of pores.

Deep plowing can reduce compaction below the plow zone.

C.O.L.E.

One of the important engineering properties of soils is related to the clay content and consistency and is called the coefficient of linear extensibility or C.O.L.E. It is a measure of the shrink-swell potential, or the volume change of the soil with changes in moisture content. If the C.O.L.E. exceeds.09, significant shrink-swell potential can be expected.

Soils with high C.O.L.E. values can cave in basements, displace walls, and break gas or water pipes. Knowing which soils have high C.O.L.E. values is an important step in selecting sites for construction activities. Compacting soils reduces the ability of soils to take on water and reduces their shrink/swell potential.

SOIL COMPACTION AND BUILDING

Soil compaction is defined as the method of mechanically increasing the density of soil. In construction, this is a significant part of the building process. If performed improperly, settlement of the soil could occur and result in unnecessary maintenance costs or structure failure. Almost all types of building sites and construction projects utilize mechanical compaction techniques Soil density is used almost exclusively by the transportation industry to specify, estimate, measure, and control soil compaction. Soil density can be easily determined via weight and volume measurements. The objective of compaction is to stabilize soils and improve their engineering properties. Source: US Dept. of Transportation

There are five principle reasons to compact soil:-

- Increases load-bearing capacity-
- Prevents soil settlement and frost damage-
- Provides stability-
- Reduces water seepage, swelling and contraction-
- Reduces settling of soil-source Soil Compaction Handbook.

Porosity

Bulk Density is an indirect measure of soil pore space. In fact, there is a formula: 100-((B.D. ÷ Particle Density) X 100)=% pore space or POROSITY. Under field conditions, pore spaces are occupied at all times by air and water. "Tortuous pathways" best describes soil pores. Porosity, when expressed as a percent, is the same thing as percent pore space. Soil particles have irregular shapes, and thus the spaces or pores between them vary irregularly in size, shape and direction. Sandy soils have large continuous pores, while clays have small pores which transmit water slowly. Clays, however, contain more pore space than sandy soils, because of the pores inside of the soil peds. To growing plants, pore sizes are of more importance than total pore space. We will be discussing this more in the unit on soil water.

Particle Density

In the laboratory you will determine porosity and particle density for the sandy soil used previously. Particle density will also be determined for this soil. Particle density only takes into account the volume occupied by the solid particles. It excludes the volume occupied by air and water. Since a large portion of most soils is composed of particles derived from quartz minerals, the particle density of most soils is near 2.6 g/cc, which is the density of quartz. Variations in the particle density are due to the presence of heavier minerals like iron oxides or lighter organic components.

Significance of Porosity

Above, a weakly developed soil is on the left and a well developed soil is on the right. Where clay accumulation is minimal, the macro pore space is about the same with increasing depth. However, where clay has moved into pores, the amount of macro pores and the rate of water movement both decrease. Soils with clay films will have slow water flow through the Bt horizon. Some soils are difficult to aerate by plowing due to the turf cover. For lawns, playgrounds, or football fields, aeration is accomplished by making small holes in the turf with solid tines or using hollow tines to pull out a small core. Later the holes can filled with sand to promote air exchange. This is called *aerifying* and *topdressing*. Many home lawns are compacted because of the construction activities when the home was built or due to people walking or riding on the turf when the soil is wet. Aerification will improve the turf growth on these lawns.

Soil Temperature Regimes

The temperature of a given soil at a given time is dependent upon the gains and losses of heat energy. Generally, dark surfaces will absorb more heat than light surfaces. However, the amount of water in the soil is an important factor. Losses of the absorbed heat are by radiation back into the atmosphere as long-wave radiation, heating the air and cooling the soil.

These soils warm quickly in the spring due to their dark surface. Soil temperature regimes are used to classify soils. They are defined according to the average annual soil temperature in the root zone. The use of soils for agriculture and forestry is closely related to soil temperature, due to the specific requirements of plants. Over most of the earth, daily soil temperatures below 50 cm deep seldom change. To approximate the mean annual soil temperature, 2 º F is added to the mean annual air temperature. If we do this for the annual temperature map of Minnesota and follow the 45ºF line, which would be 47º soil temperature, the state is divided along a line from the Twin Cities to the South Dakota border, and soils are frigid to the north and mesic to the south.

CALCULATING SOIL ORGANIC MATTER/HUMUS DECAY RATE

In general, the organic matter in the soil comes from plant and animal residues when organic matter decomposes, by the action of microorganisms and other forces. It produces substance and nutrient that helps to support plant life. Well -decomposed organic mater is called humus, which helps to hold moisture in the soil. In this study humification and mineralization are the two different combined processed used for quantifying the soil organic matter/ humus decay in a year.

What is Humification?

Humification, the transformation of organic matter into humus leads to soil decay, is a fascinating process. Organic materials such as manure or field wastes, when disked into the upper six inches of topsoil will undergo several changes. The humification process involves first catabolism, then anabolism. These are not truly correct terms as they are usually used for same functions within living organisms, but we may soil as one living organism.

What is Mineralization?

Is the process of freeing minerals from organic molecules (carbon bonds). During humification there are two possible end products for atoms within the starting molecules. Minerals may either build up stable humus or be free in form; carbon either tied within humus or evolves in form of carbon dioxide. The following is a summary table of where atoms end up after humification.

All minerals within organic compounds	80% freed up and 20% in humus
Carbon from carbohydrates	80% evolves as carbon dioxide and 20% goes to humus formation.
Nitrogen	Upto 50% will remain in humic form.

RESULTS AND DISCUSSION

The result of calculating the humus decay rate by Koef equation is classified on the basis of their severity and value ranges. Six classes have been used. Following ratings were adopted in relation to the climatic index as shown in table below. The rate of humus decay in percentage is mapped in Figure below.

Soil organic matter plays a central role in forming soil aggregates. Soils with high aggregate stability are more resistant to degradation by erosion and compaction. As the clay content of a soil increases, it requires a higher content of organic matter to maintain a given level of aggregate formation and stability. In this study it was found that, increase in clay and $CaCO_3$ content of soil correspondingly shows possession of high organic matter. Soil organic matter acts as glue due to its polysaccharide content, holding soil particles in stable aggregates.

The rate of decay varies according to the texture of the soil, being faster in a sandy than in a clay soil, the nature of the organic matter, the pH and water content above field capacity. Soil reaction between pH 5.0 and 7.5 has less effect on biological degradation. Slope aspect influences soil temperature and humidity, but slope angle has little influence on biological degradation. Altitude is taken into account by the climatic index, through its effect upon temperature.

CONCLUSION

The optimum organic matter content of soil depends on local climate, the amount and type of clay material present in the soil, and the soil, and the soil's intended use. Ideally, in a good-quality soil, all the functions of organic matter listed earlier operate fully. However, soil organic matter alone cannot supply the quantity and balance of nutrients required by crops under intensive production; fertilizer must be added to meet the crop's needs. On the other hand, soil needs optimum amounts of organic matter to maintain its structure and keep it in a tillable condition. The current study shows that the soil organic matter and humus decay rate content is an important component in assessing and maintaining the productivity of the soils. In coming days it is expected that concern of biologically degraded soil among the farmers and in soil science as a whole might a better way to make the soil healthier.

8

Agriculture and Agricultural Science

The two terms are often confused. However, they cover different concepts:

- Agriculture is the set of activities that transform the environment for the production of animals and plants for human use. Agriculture concerns techniques, including the application of agronomic research.
- Agronomy is research and development related to studying and improving plant-based agriculture.

Agricultural sciences include research and development on:

- Production techniques (e.g., irrigation management, recommended nitrogen inputs)
- Improving agricultural productivity in terms of quantity and quality (e.g., selection of drought-resistant crops and animals, development of new pesticides, yield-sensing technologies, simulation models of crop growth, in-vitro cell culture techniques)
- Transformation of primary products into end-consumer products (e.g., production, preservation, and packaging of dairy products)
- Prevention and correction of adverse

AGRICULTURAL SCIENCE: A LOCAL SCIENCE

With the exception of theoretical agronomy, research in agronomy, more than in any other field, is strongly related to local areas. It can be considered a science of ecoregions, because it is closely linked to soil properties and climate, which are never exactly the same from one place to another. Many people think an agricultural production system relying on local weather, soil characteristics, and specific crops has to be studied locally. Others feel a need to know and understand production systems in as many areas as possible, and the human dimension of interaction with nature.

History of Agricultural Science

Agricultural science began with Gregor Mendel's genetic work, but in modern terms might be better dated from the chemical fertilizer outputs of plant physiological understanding in eighteenth century Germany. In the United States, a scientific revolution in agriculture began with the Hatch Act of 1887, which used the term "agricultural science". The Hatch Act was driven by farmers' interest in knowing the constituents of early artificial fertilizer. The Smith-Hughes Act of 1917 shifted agricultural education back to its vocational roots, but the scientific foundation had been built. After 1906, public expenditures on agricultural research in the US exceeded private expenditures for the next 44 years. Intensification of agriculture since the 1960s in developed and developing countries, often referred to as the Green Revolution, was closely tied to progress made in selecting and improving crops and animals for high productivity, as well as to developing additional inputs such as artificial fertilizers and phytosanitary products. As the oldest and largest human intervention in nature, the environmental impact of agriculture in general and more recently intensive agriculture, industrial development, and population growth have raised many questions among agricultural scientists and have led to the development and emergence of new fields. These include technological fields that assume the solution to technological problems lies in better technology, such as integrated pest management, waste treatment technologies, landscape architecture, genomics, and agricultural philosophy fields that include references to food production as something essentially different from non-essential economic 'goods'. In fact, the interaction between these two approaches provide a fertile field for deeper understanding in agricultural science.

New technologies, such as biotechnology and computer science (for data processing and storage), and technological advances have made it possible to develop new research fields, including genetic engineering, agrophysics, improved statistical analysis, and precision farming. Balancing these, as above, are the natural and human sciences of agricultural science that seek to understand the human-nature interactions of traditional agriculture, including interaction of religion and agriculture, and the non-material components of agricultural production systems.

Prominent Agricultural Scientists

- Justus von Liebig
- Jethro Tull
- Robert Bakewell

- Norman Borlaug
- Luther Burbank
- George Washington Carver
- René Dumont
- Jay Lush
- Gregor Mendel
- Kailas Nath Kaul
- Louis Pasteur
- Eli Whitney
- Sewall Wright
- Scott Hill

Agricultural Science and Agriculture Crisis

Agriculture sciences seek to feed the world's population while preventing biosafety problems that may affect human health and the environment. This requires promoting good management of natural resources and respect for the environment, and increasingly concern for the psychological wellbeing of all concerned in the food production and consumption system. Economic, environmental, and social aspects of agriculture sciences are subjects of ongoing debate. Recent crises (such as avian influenza, mad cow disease and issues such as the use of genetically modified organisms) illustrate the complexity and importance of this debate.

Agriculture

Agriculture is the cultivation of animals, plants, fungi and other life forms for food, fiber, and other products used to sustain life. Agriculture was the key implement in the rise of sedentary human civilization, whereby farming of domesticated species created food surpluses that nurtured the development of civilization. The study of agriculture is known as agricultural science. Agriculture is also observed in certain species of ant and termite, but generally speaking refers to human activities. The history of agriculture dates back thousands of years, and its development has been driven and defined by greatly different climates, cultures, and technologies. However, all farming generally relies on techniques to expand and maintain the lands suitable for raising domesticated species. For plants, this usually requires some form of irrigation, although there are methods of dryland farming; pastoral herding on rangeland is still the most common means of raising livestock. In the developed world, industrial agriculture based on large-scale monoculture has become the dominant system of modern farming, although there is growing support for sustainable agriculture (e.g. permaculture or organic agriculture). Modern agronomy, plant breeding, pesticides and fertilizers, and technological improvements have sharply increased yields from cultivation, but at the same time have caused widespread ecological damage and negative human health effects. Selective breeding and modern practices in animal husbandry such as intensive pig farming have similarly increased the output of meat, but have raised concerns about animal cruelty and the health effects of the antibiotics, growth hormones, and other chemicals commonly used in industrial meat production.

The major agricultural products can be broadly grouped into foods, fibers, fuels, and raw materials. In the 21st century, plants have been used to grow biofuels, biopharmaceuticals, bioplastics, and pharmaceuticals. Specific foods include cereals, vegetables, fruits, and meat. Fibers include cotton, wool, hemp, silk and flax. Raw materials include lumber and bamboo. Other useful materials are produced by plants, such as resins. Biofuels include methane from biomass, ethanol, and biodiesel. Cut flowers, nursery plants, tropical fish and birds for the pet trade are some of the ornamental products. In 2007, one third of the world's workers were employed in agriculture. The services sector has overtaken agriculture as the economic sector employing the most people worldwide. Despite the size of its workforce, agricultural production accounts for less than five percent of the gross world product (an aggregate of all gross domestic products).

Etymology

The word *agriculture* is the English adaptation of Latin *agricultûra,* from *ager,* "a field", and *cultûra,* "cultivation" in the strict sense of "tillage of the soil". Thus, a literal reading of the word yields "tillage of a field / of fields".

Overview

Agricultural practices such as irrigation, crop rotation, fertilizers, and pesticides were developed long ago, but have made great strides in the past century. The history of agriculture has played a major role in human history, as agricultural progress has been a crucial factor in worldwide socio-economic change. Division of labor in agricultural societies made commonplace specializations rarely seen in hunter-gatherer cultures. So, too, are arts such as epic literature and monumental architecture, as well as codified legal systems. When farmers became capable of producing food beyond the needs of their own families, others in their society were freed to devote themselves to projects other than food acquisition. Historians and anthropologists have long argued that the development of agriculture made civilization possible. The total world population probably never exceeded 15 million inhabitants before the invention of agriculture.

Ancient Origins

The Fertile Crescent of Western Asia, Egypt, and India were sites of the earliest planned sowing and harvesting of plants that had previously been gathered in the wild. Independent development of agriculture occurred in northern and southern China, Africa's Sahel, New Guinea and several regions of the Americas. The eight so-called Neolithic founder crops of agriculture appear: first emmer wheat and einkorn wheat, then hulled barley, peas, lentils, bitter vetch, chick peas and flax.

By 7000 BC, small-scale agriculture reached Egypt. From at least 7000 BC the Indian subcontinent saw farming of wheat and barley, as attested by archaeological excavation at Mehrgarh in Balochistan in what is present day Pakistan. By 6000 BC, mid-scale farming was entrenched on the banks of the Nile. This, as irrigation had not yet matured sufficiently. About this time, agriculture was developed independently in the Far East, with rice, rather than wheat, as the primary crop. Chinese and Indonesian farmers went on to domesticate taro and beans including mung, soy and azuki. To complement these new sources of carbohydrates, highly organized net fishing of rivers, lakes and ocean shores in these areas brought in great volumes of essential protein. Collectively, these new methods of farming and fishing inaugurated a human population boom that dwarfed all previous expansions and continues today.

By 5000 BC, the Sumerians had developed core agricultural techniques including large-scale intensive cultivation of land, monocropping, organized irrigation, and the use of a specialized labor force, particularly along the waterway now known as the Shatt al-Arab, from its Persian Gulf delta to the confluence of the Tigris and Euphrates. Domestication of wild aurochs and mouflon into cattle and sheep, respectively, ushered in the large-scale use of animals for food/fiber and as beasts of burden. The shepherd joined the farmer as an essential provider for sedentary and seminomadic societies. Maize, manioc, and arrowroot were first domesticated in the Americas as far back as 5200 BC.

The potato, tomato, pepper, squash, several varieties of bean, tobacco, and several other plants were also developed in the Americas, as was extensive terracing of steep hillsides in much of Andean South America. The Greeks and Romans built on techniques pioneered by the Sumerians, but made few fundamentally new advances. Southern Greeks struggled with very poor soils, yet managed to become a dominant society for years. The Romans were noted for an emphasis on the cultivation of crops for trade.

In the same region, a parallel agricultural revolution occurred, resulting in some of the most important crops grown today. In Mesoamerica wild teosinte was transformed through human selection into the ancestor of modern maize, more than 6000 years ago. It gradually spread across North America and was the major crop of Native Americans at the time of European exploration. Other Mesoamerican crops include hundreds of varieties of squash and beans. Cocoa was also a major crop in domesticated Mexico and Central America. The turkey, one of the most important meat birds, was probably domesticated in Mexico or the U.S. Southwest. In the Andes region of South America the major domesticated crop was potatoes, domesticated perhaps 5000 years ago. Large varieties of beans were domesticated, in South America, as well as animals, including llamas, alpacas, and guinea pigs. Coca, still a major crop, was also domesticated in the Andes.

A minor center of domestication, the indigenous people of the Eastern U.S. appear to have domesticated numerous crops. Sunflowers, tobacco, varieties of squash and Chenopodium, as well as crops no longer grown, including marshelder and little barley were domesticated. Other wild foods may have undergone some selective cultivation, including wild rice and maple sugar. The most common varieties of strawberry were domesticated from Eastern North America.

By 3500 BC, the simplest form of the plough was developed, called the ard. Before this period, simple digging sticks or hoes were used. These tools would have also been easier to transport, which was a benefit as people only stayed until the soil's nutrients were depleted. However, through excavations in Mexico it has been found that the continuous cultivating of smaller pieces of land would also have been a sustaining practice. Additional research in central Europe later revealed that agriculture was indeed practiced at this method. For this method, ards were thus much more efficient than digging sticks.

Middle Ages

During the Middle Ages, farmers in North Africa, the Near East, and Europe began making use of agricultural technologies including irrigation systems based on hydraulic and hydrostatic principles, machines such as norias, water-raising machines, dams, and reservoirs. This combined with the invention of a three-field system of crop rotation and the moldboard plow greatly improved agricultural efficiency.

In the European medieval period, agriculture was considered part of the set of *seven mechanical arts.*

Modern Era

After 1492, a global exchange of previously local crops and livestock breeds occurred. Key crops involved in this exchange included the tomato, maize, potato, manioc, cocoa bean and tobacco going from the New World to the Old, and several varieties of wheat, spices, coffee, and sugar cane going from the Old World to the New. The most important animal exportation from the Old World to the New were those of the horse and dog (dogs were already present in the pre-Columbian Americas but not in the numbers and breeds suited to farm work).

Although not usually food animals, the horse (including donkeys and ponies) and dog quickly filled essential production roles on western-hemisphere farms.

The potato became an important staple crop in northern Europe. Since being introduced by Portuguese in the 16th century, maize and manioc have replaced traditional African crops as the continent's most important staple food crops.

By the early 19th century, agricultural techniques, implements, seed stocks and cultivar had so improved that yield per land unit was many times that seen in the Middle Ages.

Although there is a vast and interesting history of crop cultivation before the dawn of the 20th century, there is little question that the work of Charles Darwin and Gregor Mendel created the scientific foundation for plant breeding that led to its explosive impact over the past 150 years.

With the rapid rise of mechanization in the late 19th century and the 20th century, particularly in the form of the tractor, farming tasks could be done with a speed and on a scale previously impossible. These advances have led to efficiencies enabling certain modern farms in the United States, Argentina, Israel, Germany, and a few other nations to output volumes of high-quality produce per land unit at what may be the practical limit. The Haber-Bosch method for synthesizing ammonium nitrate represented a major breakthrough and allowed crop yields to overcome previous constraints. In the past century agriculture has been characterized by enhanced productivity, the substitution of labor for synthetic fertilizers and pesticides, water pollution, and farm subsidies. In recent years there has been a backlash against the external environmental effects of conventional agriculture, resulting in the organic movement.

The cereals rice, corn, and wheat provide 60% of human food supply. Between 1700 and 1980, "the total area of cultivated land worldwide increased 466%" and yields increased dramatically, particularly because of selectively bred high-yielding varieties, fertilizers, pesticides, irrigation, and machinery. For example, irrigation increased corn yields in eastern Colorado by 400 to 500% from 1940 to 1997.

However, concerns have been raised over the sustainability of intensive agriculture. Intensive agriculture has become associated with decreased soil quality in India and Asia, and there has been increased concern over the effects of fertilizers and pesticides on the environment, particularly as population increases and food demand expands. The monocultures typically used in intensive agriculture increase the number of pests, which are controlled through pesticides. Integrated pest management (IPM), which "has been promoted for decades and has had some notable successes" has not significantly affected the use of pesticides because policies encourage the use of pesticides and IPM is knowledge-intensive.

Although the "Green Revolution" significantly increased rice yields in Asia, yield increases have not occurred in the past 15–20 years. The genetic "yield potential" has increased for wheat, but the yield potential for rice has not increased since 1966, and the yield potential for maize has "barely increased in 35 years". It takes a decade or two for herbicide-resistant weeds to emerge, and insects become resistant to insecticides within about a decade. Crop rotation helps to prevent resistances.

Agricultural exploration expeditions, since the late 19th century, have been mounted to find new species and new agricultural practices in different areas of the world. Two early examples of expeditions include Frank N. Meyer's fruit- and nut-collecting trip to China and Japan from 1916-1918 and the Dorsett-Morse Oriental Agricultural Exploration Expedition to China, Japan, and Korea from 1929-1931 to collect soybean germplasm to support the rise in soybean agriculture in the United States.

In 2009, the agricultural output of China was the largest in the world, followed by the European Union, India and the United States, according to the International Monetary Fund. Economists measure the total factor productivity of agriculture and by this measure agriculture in the United States is roughly 2.6 times more productive than it was in 1948.

Six countries - the US, Canada, France, Australia, Argentina and Thailand - supply 90% of grain exports. The United States controls almost half of world grain exports. Water deficits, which are already spurring heavy grain imports in numerous middle-sized countries, including Algeria, Iran, Egypt, and Mexico, may soon do the same in larger countries, such as China or India.

FOOD SAFETY, LABELING AND REGULATION

Food security issues also coincide with food safety and food labeling concerns. Currently a global treaty, the BioSafety Protocol, regulates the trade of GMOs. The EU currently requires all GMO foods to be labeled, whereas the US does not require transparent labeling of GMO foods. Since there are still questions regarding the safety and risks associated with GMO foods, some believe the public should have the freedom to choose and know what they are eating and require all GMO products to be labeled. The Food and Agriculture Organization of the United Nations (FAO) leads international efforts to defeat hunger and provides a neutral forum where nations meet as equals to negotiate agreements and debate food policy and the regulation of agriculture. According to Dr. Samuel Jutzi, director of FAO's animal production and health division, lobbying by "powerful" big food corporations has stopped reforms that would improve human health and the environment. The "real, true issues are not being addressed by the political process because of the influence of lobbyists, of the true powerful entities," he said, speaking at the Compassion in World Farming annual forum. For example, recent proposals for a voluntary code of conduct for the livestock industry that would have provided incentives for improving standards for health, and environmental regulations, such as the number of animals an area of land can support without long-term damage, were successfully defeated due to large food company pressure.

Environmental Impact

Agriculture imposes external costs upon society through pesticides, nutrient runoff, excessive water usage, and assorted other problems. A 2000 assessment of agriculture in the UK determined total external costs for 1996 of £2,343 million, or £208 per hectare. A 2005 analysis of these costs in the USA concluded that cropland imposes approximately $5 to 16 billion ($30 to $96 per hectare), while livestock production imposes $714 million. Both studies concluded that more should be done to internalize external costs, and neither included subsidies in their analysis, but noted that subsidies also influence the cost of agriculture to society. Both focused on purely fiscal impacts. The

2000 review included reported pesticide poisonings but did not include speculative chronic effects of pesticides, and the 2004 review relied on a 1992 estimate of the total impact of pesticides.

Agriculture accounts for 70 per cent of withdrawals of freshwater resources. However, increasing pressure being placed on water resources by industry, cities and the involving biofuels industry means that water scarcity is increasing and agriculture is facing the challenge of producing more food for the world's growing population with fewer water resources. Scientists are also realizing that water resources need to be allocated to maintain natural environmental services, such as protecting towns from flooding, cleaning ecosystems and supporting fish stocks.

In the book *Out of Water: From abundance to scarcity and how to solve the world's water problems,* authors Colin Chartres and Samyukta Varma of the International Water Management Institute lay down a six-point plan of action for addressing the global challenge of producing sufficient food for the world with dwindling water resources. One of the actions they say is required is to ensure all water systems, such as lakes and rivers, have water allocated to environmental flow. A key player who is credited to saving billions of lives because of his revolutionary work in developing new agricultural techniques is Norman Borlaug. His transformative work brought high-yield crop varieties to developing countries and earned him an unofficial title as the father of the Green Revolution.

Livestock Issues

A senior UN official and co-author of a UN report detailing this problem, Henning Steinfeld, said "Livestock are one of the most significant contributors to today's most serious environmental problems". Livestock production occupies 70% of all land used for agriculture, or 30% of the land surface of the planet. It is one of the largest sources of greenhouse gases, responsible for 18% of the world's greenhouse gas emissions as measured in CO_2 equivalents. By comparison, all transportation emits 13.5% of the CO_2. It produces 65% of human-related nitrous oxide (which has 296 times the global warming potential of CO_{2}) and 37% of all human-induced methane (which is 23 times as warming as CO_2. It also generates 64% of the ammonia, which contributes to acid rain and acidification of ecosystems. Livestock expansion is cited as a key factor driving deforestation, in the Amazon basin 70% of previously forested area is now occupied by pastures and the remainder used for feedcrops. Through deforestation and land degradation, livestock is also driving reductions in biodiversity.

Land Transformation and Degradation

Land transformation, the use of land to yield goods and services, is the most substantial way humans alter the Earth's ecosystems, and is considered the driving force in the loss of biodiversity. Estimates of the amount of land transformed by humans vary from 39–50%. Land degradation, the long-term decline in ecosystem function and productivity, is estimated to be occurring on 24% of land worldwide, with cropland overrepresented. The UN-FAO report cites land management as the driving factor behind degradation and reports that 1.5 billion people rely upon the degrading land. Degradation can be deforestation, desertification, soil erosion, mineral depletion, or chemical degradation (acidification and salinization).

Eutrophication

Eutrophication, excessive nutrients in aquatic ecosystems resulting in algal blooms and anoxia, leads to fish kills, loss of biodiversity, and renders water unfit for drinking and other industrial uses. Excessive fertilization and manure application to cropland, as well as high livestock stocking densities cause nutrient (mainly nitrogen and phosphorus) runoff and leaching from agricultural land. These nutrients are major nonpoint pollutants contributing to eutrophication of aquatic ecosystems.

Pesticides

Pesticide use has increased since 1950 to 2.5 million tons annually worldwide, yet crop loss from pests has remained relatively constant. The World Health Organization estimated in 1992 that 3 million pesticide poisonings occur annually, causing 220,000 deaths. Pesticides select for pesticide resistance in the pest population, leading to a condition termed the 'pesticide treadmill' in which pest resistance warrants the development of a new pesticide.

An alternative argument is that the way to 'save the environment' and prevent famine is by using pesticides and intensive high yield farming, a view exemplified by a quote heading the Center for Global Food Issues website: 'Growing more per acre leaves more land for nature'. However, critics argue that a trade-off between the environment and a need for food is not inevitable, and that pesticides simply replace good agronomic practices such as crop rotation.

Climate Change

Climate change has the potential to affect agriculture through changes in temperature, rainfall (timing and quantity), CO_2, solar radiation and the interaction of these elements. Agriculture can both mitigate or worsen global warming. Some of the increase in CO_2 in the atmosphere comes from the decomposition of organic matter in the soil, and much of the methane emitted into the atmosphere is caused by the decomposition of organic matter in wet soils such as rice paddies. Further, wet or anaerobic soils also lose nitrogen through denitrification, releasing the greenhouse gas nitric oxide. Changes in management can reduce the release of these greenhouse gases, and soil can further be used to sequester some of the CO_2 in the atmosphere.

International Economics and Agriculture

Differences in economic development, population density and culture mean that the farmers of the world operate under very different conditions. A US cotton farmer may receive US$230 in government subsidies per acre planted (in 2003), while farmers in Mali and other third-world countries do without. When prices decline, the heavily subsidized US farmer is not forced to reduce his output, making it difficult for cotton prices to rebound, but his Mali counterpart may go broke in the meantime. A livestock farmer in South Korea can calculate with a (highly subsidized) sales price of US$1300 for a calf produced. A South American Mercosur country rancher calculates with a calf's sales price of US$120–200 (both 2008 figures). With the former, scarcity and high cost of land is compensated with public subsidies, the latter compensates absence of subsidies with economics of scale and low cost of land. In the Peoples Republic of China, a rural household's productive asset may be one hectare of farmland. In Brazil, Paraguay and other countries where local legislature allows such purchases, international investors buy thousands of hectares of farmland or raw land at prices of a few hundred US$ per hectare.

List of Countries by Agricultural Output

Below is a list of countries by agricultural output in 2009. Output is in millions of US$.

Rank	*Country*	*Output*
1	China	520,352
—	European Union	312,498
2	India	210,116
3	United States	171,075
4	Brazil	96,016
5	Japan	81,089
6	Russia	57,774
7	Spain	48,313

8	France	48,167
9	Australia	40,885
10	Italy	38,129

Energy and Agriculture

Since the 1940s, agricultural productivity has increased dramatically, due largely to the increased use of energy-intensive mechanization, fertilizers and pesticides. The vast majority of this energy input comes from fossil fuel sources. Between 1950 and 1984, the Green Revolution transformed agriculture around the globe, with world grain production increasing by 250% as world population doubled. Modern agriculture's heavy reliance on petrochemicals and mechanization has raised concerns that oil shortages could increase costs and reduce agricultural output, causing food shortages.

Agriculture and Food System Share (%) of Total Energy Consumption by Three Industrialized Nations

Country	*Year*	*Agriculture (direct & indirect)*	*Food system*
United Kingdom	2005	1.9	11
United States of America	1996	2.1	10
United States of America	2002	2.0	14
Sweden	2000	2.5	13

Modern or industrialized agriculture is dependent on fossil fuels in two fundamental ways: 1) direct consumption on the farm and 2) indirect consumption to manufacture inputs used on the farm. Direct consumption includes the use of lubricants and fuels to operate farm vehicles and machinery; and use of gas, liquid propane, and electricity to power dryers, pumps, lights, heaters, and coolers. American farms directly consumed about 1.2 exajoules (1.1 quadrillion BTU) in 2002, or just over 1 percent of the nation's total energy. Indirect consumption is mainly oil and natural gas used to manufacture fertilizers and pesticides, which accounted for 0.6 exajoules (0.6 quadrillion BTU) in 2002. The energy used to manufacture farm machinery is also a form of indirect agricultural energy consumption, but it is not included in USDA estimates of U.S. agricultural energy use. Together, direct and indirect consumption by U.S. farms accounts for about 2 percent of the nation's energy use. Direct and indirect energy consumption by U.S. farms peaked in 1979, and has gradually declined over the past 30 years. Food systems encompass not just agricultural production, but also off-farm processing, packaging, transporting, marketing, consumption, and disposal of food and food-related items. Agriculture accounts for less than one-fifth of food system energy use in the United States. Oil shortages could impact this food supply. Some farmers using modern organic-farming methods have reported yields as high as those available from conventional farming without the use of synthetic fertilizers and pesticides. However, the reconditioning of soil to restore nutrients lost during the use of monoculture agriculture techniques made possible by petroleum-based technology takes time.

In 2007, higher incentives for farmers to grow non-food biofuel crops combined with other factors (such as over-development of former farm lands, rising transportation costs, climate change, growing consumer demand in China and India, and population growth) to cause food shortages in Asia, the Middle East, Africa, and Mexico, as well as rising food prices around the globe. As of December 2007, 37 countries faced food crises, and 20 had imposed some sort of food-price controls.

Some of these shortages resulted in food riots and even deadly stampedes. The biggest fossil fuel input to agriculture is the use of natural gas as a hydrogen source for the Haber-Bosch fertilizer-creation process. Natural gas is used because it is the cheapest currently available source of hydrogen. When oil production becomes so scarce that natural gas is used as a partial stopgap replacement, and hydrogen use in transportation increases, natural gas will become much more expensive. If the Haber Process is unable to be commercialized using renewable energy (such as by electrolysis) or if other sources of hydrogen are not available to replace the Haber Process, in amounts sufficient to supply transportation and agricultural needs, this major source of fertilizer would either become extremely expensive or unavailable. This would either cause food shortages or dramatic rises in food prices.

Mitigation of Effects of Petroleum Shortages

One effect oil shortages could have on agriculture is a full return to organic agriculture. In light of peak-oil concerns, organic methods are more sustainable than contemporary practices because they use no petroleum-based pesticides, herbicides, or fertilizers. Some farmers using modern organic-farming methods have reported yields as high as those available from conventional farming. Organic farming may however be more labor-intensive and would require a shift of the workforce from urban to rural areas.

It has been suggested that rural communities might obtain fuel from the biochar and synfuel process, which uses agricultural *waste* to provide charcoal fertilizer, some fuel *and* food, instead of the normal food vs fuel debate. As the synfuel would be used on-site, the process would be more efficient and might just provide enough fuel for a new organic-agriculture fusion.

It has been suggested that some transgenic plants may some day be developed which would allow for maintaining or increasing yields while requiring fewer fossil-fuel-derived inputs than conventional crops. The possibility of success of these programmes is questioned by ecologists and economists concerned with unsustainable GMO practices such as terminator seeds, and a January 2008 report shows that GMO practices "fail to deliver environmental, social and economic benefits."

While there has been some research on sustainability using GMO crops, at least prominent multi-year attempt by Monsanto Company has been unsuccessful, though during the same period traditional breeding techniques yielded a more sustainable variety of the same crop.

Policy

Agricultural policy focuses on the goals and methods of agricultural production. At the policy level, common goals of agriculture include:

- Conservation
- Economic stability
- Environmental sustainability
- Food quality: Ensuring that the food supply is of a consistent and known quality.
- Food safety: Ensuring that the food supply is free of contamination.
- Food security: Ensuring that the food supply meets the population's needs.
- Poverty reduction

Botany

Botany, plant science(s), or plant biology is a branch of biology that involves the scientific study of plant life. Botany covers a wide range of scientific disciplines concerned with the study of plants, algae and fungi, including structure, growth, reproduction, metabolism, development, diseases, chemical properties, and evolutionary relationships among taxonomic groups. Botany began with early human efforts to identify edible, medicinal and poisonous plants, making it one of the oldest sciences. Today botanists study over 550,000 species of living organisms. The term "botany" comes from Greek âïôÜíç, meaning "pasture, grass, fodder", perhaps via the idea of a livestock keeper needing to know which plants are safe for livestock to eat.

Scope and Importance of Botany

As with other life forms in biology, plant life can be studied from different perspectives, from the molecular, genetic and biochemical level through organelles, cells, tissues, organs, individuals, plant populations, and communities of plants. At each of these levels a botanist might be concerned with the classification (taxonomy), structure (anatomy and morphology), or function (physiology) of plant life. Historically all living things were grouped as animals or plants, and botany covered all organisms not considered animals. Some organisms once included in the field of botany are no longer considered to belong to the plant kingdom – these include fungi (studied in mycology), lichens (lichenology), bacteria (bacteriology), viruses (virology) and single-celled algae, which are now grouped as part of the Protista. However, attention is still given to these groups by botanists, and fungi, lichens, bacteria and photosynthetic protists are usually covered in introductory botany courses. The study of plants is vital because they are a fundamental part of life on Earth, which generates the oxygen, food, fibres, fuel and medicine that allow humans and other life forms to exist. Through photosynthesis, plants absorb carbon dioxide, a greenhouse gas that in large amounts can affect global climate. Additionally, they prevent soil erosion and are influential in the water cycle. A good understanding of plants is crucial to the future of human societies as it allows us to:

- Produce food to feed an expanding population
- Understand fundamental life processes
- Produce medicine and materials to treat diseases and other ailments
- Understand environmental changes more clearly

Paleobotanists study ancient plants in the fossil record. It is believed that early in the Earth's history, the evolution of photosynthetic plants altered the global atmosphere of the earth, changing the ancient atmosphere by oxidation.

Human Nutrition

Virtually all foods eaten come from plants, either directly from staple foods and other fruit and vegetables, or indirectly through livestock or other animals, which rely on plants for their nutrition. Plants are the fundamental base of nearly all food chains because they use the energy from the sun and nutrients from the soil and atmosphere, converting them into a form that can be consumed and utilized by animals; this is what ecologists call the first trophic level. Botanists also study how plants produce food we can eat and how to increase yields and therefore their work is important in mankind's ability to feed the world and provide food security for future generations, for example, through plant breeding.

Botanists also study weeds, plants which are considered to be a nuisance in a particular location. Weeds are a considerable problem in agriculture, and botany provides some of the basic science used to understand how to minimize 'weed' impact in agriculture and native ecosystems.

Ethnobotany is the study of the relationships between plants and people, and when this kind of study is turned to the investigation of plant-people relationships in past times, it is referred to as archaeobotany or paleoethnobotany.

Fundamental Life Processes

Plants are convenient organisms in which fundamental life processes (like cell division and protein synthesis) can be studied, without the ethical dilemmas of studying animals or humans. The genetic laws of inheritance were discovered in this way by Gregor Mendel, who was studying the way pea shape is inherited. What Mendel learned from studying plants has had far reaching benefits outside of botany. Additionally, Barbara McClintock discovered 'jumping genes' by studying maize. These are a few examples that demonstrate how botanical research has an ongoing relevance to the understanding of fundamental biological processes.

Medicine and Materials

Many medicinal and recreational drugs, like tetrahydrocannabinol, caffeine, and nicotine come directly from the plant kingdom. Others are simple derivatives of botanical natural products; for example, aspirin is based on the pain killer salicylic acid which originally came from the bark of willow trees. As well, the narcotic analgesics such as morphine are derived from the opium poppy. There may be many novel cures for diseases provided by plants, waiting to be discovered.

Popular stimulants like coffee, chocolate, tobacco, and tea also come from plants. Most alcoholic beverages come from fermenting plants such as barley (beer), rice (sake) and grapes (wine). Plants also provide us with many natural materials, such as hemp, cotton, wood, paper, linen, vegetable oils, some types of rope, and rubber.

The production of silk would not be possible without the cultivation of the mulberry plant. Sugarcane, rapeseed, soy and other plants with a highly fermentable sugar or oil content have recently been put to use as sources of biofuels, which are important alternatives to fossil fuels.

Environmental Changes

Plants can also help us understand changes in on our environment in many ways.

- Understanding habitat destruction and species extinction is dependent on an accurate and complete catalog of plant systematics and taxonomy.
- Plant responses to ultraviolet radiation can help us monitor problems like ozone depletion.
- Analyzing pollen deposited by plants thousands or millions of years ago can help scientists to reconstruct past climates and predict future ones, an essential part of climate change research.
- Recording and analyzing the timing of plant life cycles are important parts of phenology used in climate-change research.
- Lichens, which are sensitive to atmospheric conditions, have been extensively used as pollution indicators.

In many different ways, plants can act a little like the 'miners' canary', an early warning system alerting us to important changes in our environment. In addition to these practical and scientific reasons, plants are extremely valuable as recreation for millions of people who enjoy gardening, horticultural and culinary uses of plants every day.

History

Early Botany

Early examples of plant taxonomy occur in the Rigveda, that divides plants into *V[ska* (tree), *Osadhi* (herbs useful to humans) and *Virudha* (creepers), which are then further subdivided. The Atharvaveda divides plants into eight classes, *Visakha* (spreading branches), *Manjari* (leaves with long clusters), *Sthambini* (bushy plants), *Prastanavati* (which expands); *Ekas[nga* (those with monopodial growth), *Pratanavati* (creeping plants), *Amsumati* (with many stalks), and *Kandini* (plants with knotty joints). The Taittiriya Samhita classifies the plant kingdom into *v[ksa, vana* and *druma* (trees), *visakha* (shrubs with spreading branches), *sasa* (herbs), *amsumali* (a spreading or deliquescent plant), *vratati* (climber), *stambini* (bushy plant), *pratanavati* (creeper), and *alasala* (those spreading on the ground). Manusmriti – Law book of Hindus – proposed a classification of plants in eight major categories. Charaka Samhitâ and Sushruta Samhita and the Vaisesikas also present an elaborate taxonomy. Parashara, the author of *V[ksayurveda* (the science of life of trees), classifies plants into Dvimatrka (Dicotyledons) and Ekamatrka (Monocotyledons). These are further classified into *Samiganiya* (Fabaceae), *Puplikagalniya* (Rutaceae), *Svastikaganiya*

(Cruciferae), *Tripuspaganiya* (Cucurbitaceae), *Mallikaganiya* (Apocynaceae), and *Kurcapuspaganiya* (Asteraceae). Important medieval Indian works of plant physiology include the *Prthviniraparyam* of Udayana, *Nyayavindutika* of Dharmottara, *Saddarsana-samuccaya* of Gunaratna, and *Upaskara* of Sankaramisra.

Ancient Iranic People

The knowledge of medical plants and botany was considered as secret and holy by the ancient Iranic people. There is evidence of such practices in the documents that have survived from the ancient Zoroastrian writings. The practice and use of botany for medical purposes as well as various Iranic cousins and traditions is still common to this day amongst the Iranic people of the Central Asia, Near East and Europe.

Ancient China

In ancient China, the recorded listing of different plants and herb concoctions for pharmaceutical purposes spans back to at least the Warring States (481 BC-221 BC). Many Chinese writers over the centuries contributed to the written knowledge of herbal pharmaceutics.

There was the Han Dynasty (202 BC-220 AD) written work of the Huangdi Neijing and the famous pharmacologist Zhang Zhongjing of the 2nd century. There was also the 11th century scientists and statesmen Su Song and Shen Kuo, who compiled treatises on herbal medicine and included the use of mineralogy.

Greco-Roman world

Among the earliest of botanical works in Europe, written around 300 B.C., are two large treatises by Theophrastus: *On the History of Plants* (*Historia Plantarum*) and *On the Causes of Plants*. Together these books constitute the most important contribution to botanical science during antiquity and on into the Middle Ages. Aristotle also wrote about plants. One theory that Greco-Romans came up with about plants was that they ate soil for nutrients.

The Greek medical writer Pedanius Dioscorides (ca.40-90) provides important evidence on Greek and Roman knowledge of medicinal plants. Dioscorides is famous for writing a five volume book in his native Greek Ðåñß ýëçò éáôñéêÞò (*De Materia Medica* - in the Latin translation) that is one of the most influential herbal books in history. In fact, it remained in use until about CE 1600. Approximately 1300-1400 different plant species were known under Roman reign.

Medieval Botany

The earliest known work from the Muslim world dedicated to the study of agriculture was Ibn Wahshiyya's Nabatean Agriculture, which also dealt with the related field of botany and was also an early cookbook. The Kurdish biologist Abû $anîfa Dînawarî (828-896) is considered the founder of Arabic botany for his *Book of Plants,* in which he described at least 637 plants and discussed plant development from germination to death, describing the phases of plant growth and the production of flowers and fruit. Theophrastus's *Historia Plantarum* served as a reference point in botany for many centuries, and was further developed around 1200 A.D. by Giovanni Bodeo da Stapelio, who added a *commentarius* and drawings: see Historia Plantarum —*Selected pages of a 17th century edition of the 1200 A.D. version (in Italian).* Ibn Bassal is known for his famous work named *The Classification of Soils.* Al-Asma'i was the earliest known Arab biologist, botanist and zoologist. al-Masihi was the first to recognize the science of Botany. In the early 13th century, the Andalusian-Arabian biologist Abu al-Abbas al-Nabati developed an early scientific method for botany, introducing empirical and experimental techniques in the testing, description and identification of numerous materia medica, and separating unverified reports from those supported by actual tests and observations. His student Ibn al-Baitar (d. 1248) wrote a pharmaceutical encyclopedia describing 1,400 plants, foods, and drugs, 300 of which were his own original discoveries. A Latin translation of his work was useful to European biologists and pharmacists in the 18th and 19th centuries.

Early Modern Botany

German physician Leonhart Fuchs (1501–1566) was one of the three founding fathers of botany, along with Otto Brunfels (1489- 1534) and Hieronymus Bock (1498–1554) (also called Hieronymus Tragus). Valerius Cordus (1515–1554) authored one of the greatest pharmacopoeias and one of the most celebrated herbals in history, *Dispensatorium* (1546). As early as the 16th century, the Italian Ulisse Aldrovandi was scientifically researching plants. In 1665, using an early microscope, Robert Hooke discovered cells in cork, and a short time later in living plant tissue. The Germans Jacob Theodor Klein and Leonhart Fuchs, the Swiss Conrad von Gesner, and the British author Nicholas Culpeper published herbals that gave information on the medicinal uses of plants. During the 18th century systems of classification became deliberately artificial and served only for the purpose of identification. These classifications are comparable to diagnostic keys, where taxa are artificially grouped in pairs by few, easily recognisable characters. The sequence of the taxa in keys is often totally unrelated to their natural or phyletic groupings. In the 18th century an increasing number of new plants had arrived in Europe, from newly discovered countries and the European colonies worldwide, and a larger amount of plants became available for study. In 1754 Carl von Linné (Carl Linnaeus) divided the plant Kingdom into 25 classes. One, the*Cryptogamia,* included all the plants with concealed reproductive parts (algae, fungi, mosses and liverworts and ferns).

The increased knowledge on anatomy, morphology and life cycles, lead to the realization that there were more natural affinities between plants, than the sexual system of Linnaeus indicated. Adanson (1763), de Jussieu (1789), and Candolle (1819) all proposed various alternative natural systems that were widely followed. The ideas of natural selection as a mechanism for evolution required adaptations to the Candollean system, which started the studies on evolutionary relationships and phylogenetic classifications of plants. Botany was greatly stimulated by the appearance of the first "modern" text book, Matthias Schleiden's *Grundzuge der Wissenschaftlichen,* published in English in 1849 as *Principles of Scientific Botany.* Carl Willdenow examined the connection between seed dispersal and distribution, the nature of plant associations and the impact of geological history. The cell nucleus was discovered by Robert Brown in 1831.

Modern Botany

A considerable amount of new knowledge today is being generated from studying model plants like *Arabidopsis thaliana.* This weedy species in the mustard family was one of the first plants to have its genome sequenced. The sequencing of the rice (*Oryza sativa*) genome, its relatively small genome, and a large international research community have made rice an important cereal/ grass/monocot model. Another grass species, *Brachypodium distachyon* is also emerging as an experimental model for understanding the genetic, cellular and molecular biology of temperate grasses. Other commercially important staple foods like wheat, maize, barley, rye, pearl millet and soybean are also having their genomes sequenced. Some of these are challenging to sequence because they have more than two haploid (n) sets of chromosomes, a condition known as polyploidy, common in the plant kingdom. *Chlamydomonas reinhardtii* (a single-celled, green alga) is another plant model organism that has been extensively studied and provided important insights into cell biology.

In 1998 the Angiosperm Phylogeny Group published a phylogeny of flowering plants based on an analysis of DNA sequences from most families of flowering plants. As a result of this work, major questions such as which families represent the earliest branches in the genealogy of angiosperms are now understood. Investigating how plant species are related to each other allows botanists to better understand the process of evolution in plants.

Horticulture

Horticulture is the industry and science of plant cultivation including the process of preparing soil for the planting of seeds, tubers, or cuttings. Horticulturists work and conduct research in the disciplines of plant propagation and cultivation, crop production, plant breeding and genetic engineering, plant biochemistry, and plant physiology. The work involves fruits, berries, nuts, vegetables, flowers, trees, shrubs, and turf. Horticulturists work to improve crop yield, quality, nutritional value, and resistance to insects, diseases, and environmental stresses. Horticulture usually refers to gardening on a smaller scale, while agriculture refers to the large-scale cultivation of crops.

Etymology

The word *horticulture* is modeled after *agriculture*, and comes from the Latin *hortus* "garden" and *cultûra* "cultivation", from *cultus*, the perfect passive participle of the verb *colô* "I cultivate".

Areas of Study

Horticulture involves eight areas of study, which can be grouped into two broad sections - ornamentals and edibles:

- Arboriculture is the study of, and the selection, planting, care, and removal of, individual trees, shrubs, vines, and other perennial woody plants.
- Floriculture includes the production and marketing of floral crops.
- Landscape horticulture includes the production, marketing and maintenance of landscape plants.
- Olericulture includes the production and marketing of vegetables.
- Pomology includes the production and marketing of fruits.
- Viticulture includes the production and marketing of grapes.
- Oenology includes all aspects of wine and winemaking.
- Postharvest physiology involves maintaining the quality of and preventing the spoilage of horticultural crops.

Horticulturists can work in industry, government or educational institutions or private collections. They can be cropping systems engineers, wholesale or retail business managers, propagators and tissue culture specialists (fruits, vegetables, ornamentals, and turf), crop inspectors, crop production advisers, extension specialists, plant breeders, research scientists, and of course, teachers. Disciplines which complement horticulture include biology, botany, entomology, chemistry, mathematics, genetics, physiology, statistics, computer science, and communications, garden design, planting design. Plant science and horticulture courses include: plant materials, plant propagation, tissue culture, crop production, post-harvest handling, plant breeding, pollination management, crop nutrition, entomology, plant pathology, economics, and business. Some careers in horticultural science require a masters (MS) or doctoral (PhD) degree. Horticulture is practiced in many gardens, "plant growth centres" and nurseries. Activities in nurseries range from preparing seeds and cuttings to growing fully mature plants. These are often sold or transferred to ornamental gardens or market gardens.

Anthropology

Horticulture has a very long history. The study and science of horticulture dates all the way back to the times of Alexander the Great, and has been going on ever since, with present day horticulturists such as Freeman S. Howlett, the revolutionary horticulturist. The origins of horticulture lie in the transition of human communities from nomadic hunter-gatherers to sedentary or semi-sedentary horticultural communities, cultivating a variety of crops on a small scale around their dwellings or in specialized plots visited occasionally during migrations from one area to the next (such as the "milpa" or maize field of Mesoamerican cultures). In the Pre-Columbian Amazon Rainforest, natives are believed to have used biochar to enhance soil productivity by smoldering plant waste. European settlers called it *Terra Preta de Indio*. In forest areas such horticulture is often carried out in swiddens ("slash and burn" areas). A characteristic of horticultural communities is that useful trees are often to be found planted around communities or specially retained from the natural ecosystem. Horticulture primarily differs from agriculture in two ways, firstly it generally encompasses a smaller scale of cultivation, using small plots of mixed crops rather than large fields of single crops. Secondly horticultural cultivations generally include a wide variety of crops, even including fruit trees with ground crops. Agricultural cultivations however as a rule focus on one primary crop. In pre-contact North America the semi-sedentary horticultural communities of the Eastern Woodlands (growing maize, squash and sunflower) contrasted markedly with the mobile hunter-gatherer communities of the Plains people. In Central America, Maya horticulture involved augmentation of the forest with useful trees such as papaya, avocado, cacao, ceiba and sapodilla.

In the cornfields, multiple crops were grown such as beans (using cornstalks as supports), squash, pumpkins and chilli peppers, in some cultures tended mainly or exclusively by women.

Plant Breeding

Plant breeding is the art and science of changing the genetics of plants for the benefit of mankind. Plant breeding can be accomplished through many different techniques ranging from simply selecting plants with desirable characteristics for propagation, to more complex molecular techniques

Plant breeding has been practiced for thousands of years, since near the beginning of human civilization. It is now practiced worldwide by individuals such as gardeners and farmers, or by professional plant breeders employed by organizations such as government institutions, universities, crop-specific industry associations or research centers.

International development agencies believe that breeding new crops is important for ensuring food security by developing new varieties that are higher-yielding, resistant to pests and diseases, drought-resistant or regionally adapted to different environments and growing conditions.

Domestication

Plant breeding in certain situations may lead to the domestication of wild plants. Domestication of plants is an artificial selection process conducted by humans to produce plants that have more desirable traits than wild plants, and which renders them dependent on artificial (usually enhanced) environments for their continued existence. The practice is estimated to date back 9,000-11,000 years. Many crops in present day cultivation are the result of domestication in ancient times, about 5,000 years ago in the Old World and 3,000 years ago in the New World. In the Neolithic period, domestication took a minimum of 1,000 years and a maximum of 7,000 years. Today, all of our principal food crops come from domesticated varieties. Almost all the domesticated plants used today for food and agriculture were domesticated in the centers of origin. In these centers there is still a great diversity of closely related wild plants, so-called crop wild relatives, that can also be used for improving modern cultivars by plant breeding.

A plant whose origin or selection is due primarily to intentional human activity is called a cultigen, and a cultivated crop species that has evolved from wild populations due to selective pressures from traditional farmers is called a landrace. Landraces, which can be the result of natural forces or domestication, are plants (or animals) that are ideally suited to a particular region or environment. An example are the landraces of rice, *Oryza sativa* subspecies *indica,* which was developed in South Asia, and *Oryza sativa* subspecies *japonica,* which was developed in China.

CLASSICAL PLANT BREEDING

Classical plant breeding uses deliberate interbreeding (crossing) of closely or distantly related individuals to produce new crop varieties or lines with desirable properties. Plants are crossbred to introduce traits/genes from one variety or line into a new genetic background. For example, a mildew-resistant pea may be crossed with a high-yielding but susceptible pea, the goal of the cross being to introduce mildew resistance without losing the high-yield characteristics. Progeny from the cross would then be crossed with the high-yielding parent to ensure that the progeny were most like the high-yielding parent, (backcrossing). The progeny from that cross would then be tested for yield and mildew resistance and high-yielding resistant plants would be further developed. Plants may also be crossed with themselves to produce inbred varieties for breeding.

Classical breeding relies largely on homologous recombination between chromosomes to generate genetic diversity. The classical plant breeder may also makes use of a number of *in vitro* techniques such as protoplast fusion, embryo rescue or mutagenesis to generate diversity and produce hybrid plants that would not exist in nature.

Traits that breeders have tried to incorporate into crop plants in the last 100 years include:

1. Increased quality and yield of the crop
2. Increased tolerance of environmental pressures (salinity, extreme temperature, drought)
3. Resistance to viruses, fungi and bacteria
4. Increased tolerance to insect pests
5. Increased tolerance of herbicides

Before World War II

Intraspecific hybridization within a plant species was demonstrated by Charles Darwin and Gregor Mendel, and was further developed by geneticists and plant breeders. In the United Kingdom in the 1880s, it was the pioneering work of Gartons Agricultural Plant Breeders. In the early 20th century, plant breeders realized that Mendel's findings on the non-random nature of inheritance could be applied to seedling populations produced through deliberate pollinations to predict the frequencies of different types. From 1904 to World War II in Italy Nazareno Strampelli created a number of wheat hybrids. His work allowed Italy to increase hugely crop production during the so called "Battle for Grain" (1925–1940) and some varieties was exported in foreign countries, as Argentina, Mexico, China and others. After the war, the work of Strampelli was quickly forgotten, but thanks to the hybrids he created, Norman Borlaug was able to move the very first steps of the Green Revolution.

In 1908, George Harrison Shull described heterosis, also known as hybrid vigor. Heterosis describes the tendency of the progeny of a specific cross to outperform both parents. The detection of the usefulness of heterosis for plant breeding has led to the development of inbred lines that reveal a heterotic yield advantage when they are crossed. Maize was the first species where heterosis was widely used to produce hybrids.

By the 1920s, statistical methods were developed to analyze gene action and distinguish heritable variation from variation caused by environment. In 1933, another important breeding technique, cytoplasmic male sterility (CMS), developed in maize, was described by Marcus Morton Rhoades. CMS is a maternally inherited trait that makes the plant produce sterile pollen. This enables the production of hybrids without the need for labor intensive detasseling.

These early breeding techniques resulted in large yield increase in the United States in the early 20th century. Similar yield increases were not produced elsewhere until after World War II, the Green Revolution increased crop production in the developing world in the 1960s.

After World War II

Following World War II a number of techniques were developed that allowed plant breeders to hybridize distantly related species, and artificially induce genetic diversity.

When distantly related species are crossed, plant breeders make use of a number of plant tissue culture techniques to produce progeny from otherwise fruitless mating. Interspecific and intergeneric hybrids are produced from a cross of related species or genera that do not normally sexually reproduce with each other. These crosses are referred to as *Wide crosses*. For example, the cereal triticale is a wheat and rye hybrid. The cells in the plants derived from the first generation created from the cross contained an uneven number of chromosomes and as result was sterile. The cell division inhibitor colchicine was used to double the number of chromosomes in the cell and thus allow the production of a fertile line.

Failure to produce a hybrid may be due to pre- or post-fertilization incompatibility. If fertilization is possible between two species or genera, the hybrid embryo may abort before maturation. If this does occur the embryo resulting from an interspecific or intergeneric cross can sometimes be rescued and cultured to produce a whole plant. Such a method is referred to as *Embryo Rescue*. This technique has been used to produce new rice for Africa, an interspecific cross of Asian rice *(Oryza sativa)* and African rice *(Oryza glaberrima).*

Hybrids may also be produced by a technique called protoplast fusion. In this case protoplasts are fused, usually in an electric field. Viable recombinants can be regenerated in culture.

Chemical mutagens like EMS and DMS, radiation and transposons are used to generate mutants with desirable traits to be bred with other cultivars - a process known as *Mutation Breeding*. Classical plant breeders also generate genetic diversity within a species by exploiting a process called somaclonal variation, which occurs in plants produced from tissue culture, particularly plants derived from callus. Induced polyploidy, and the addition or removal of chromosomes using a technique called chromosome engineering may also be used.

When a desirable trait has been bred into a species, a number of crosses to the favoured parent are made to make the new plant as similar to the favoured parent as possible. Returning to the example of the mildew resistant pea being crossed with a high-yielding but susceptible pea, to make the mildew resistant progeny of the cross most like the high-yielding parent, the progeny will be crossed back to that parent for several generations. This process removes most of the genetic contribution of the mildew resistant parent. Classical breeding is therefore a cyclical process.

With classical breeding techniques, the breeder does not know exactly what genes have been introduced to the new cultivars. Some scientists therefore argue that plants produced by classical breeding methods should undergo the same safety testing regime as genetically modified plants. There have been instances where plants bred using classical techniques have been unsuitable for human consumption, for example the poison solanine was unintentionally increased to unacceptable levels in certain varieties of potato through plant breeding. New potato varieties are often screened for solanine levels before reaching the marketplace.

Modern Plant Breeding

Modern plant breeding uses techniques of molecular biology to select, or in the case of genetic modification, to insert, desirable traits into plants.

Steps of Plant Breeding

The following are the major activities of plant breeding;

1. Creation of variation
2. Selection
3. Evaluation
4. Release
5. Multiplication
6. Distribution of the new variety

Marker Assisted Selection

Sometimes many different genes can influence a desirable trait in plant breeding. The use of tools such as molecular markers or DNA fingerprinting can map thousands of genes. This allows plant breeders to screen large populations of plants for those that possess the trait of interest. The screening is based on the presence or absence of a certain gene as determined by laboratory procedures, rather than on the visual identification of the expressed trait in the plant.

Reverse Breeding and Doubled Haploids (DH)

A method for efficiently producing homozygous plants from a heterozygous starting plant, which has all desirable traits. This starting plant is induced to produce doubled haploid from haploid cells, and later on creating homozygous/doubled haploid plants from those cells. While in natural offspring genetic recombination occurs and traits can be unlinked from each other, in doubled haploid cells and in the resulting DH plants recombination is no longer an issue.

There, a recombination between two corresponding chromosomes does not lead to un-linkage of alleles or traits, since it just leads to recombination with its identical copy. Thus, traits on one chromosome stay linked. Selecting those offspring having the desired set of chromosomes and crossing them will result in a final F1 hybrid plant, having exactly the same set of chromosomes, genes and traits as the starting hybrid plant.

The homozygous parental lines can reconstitute the original heterozygous plant by crossing, if desired even in a large quantity. An individual heterozygous plant can be converted into a heterozygous variety (F1 hybrid) without the necessity of vegetative propagation but as the result of the cross of two homozygous/doubled haploid lines derived from the originally selected plant. patent

Genetic Modification

Genetic modification of plants is achieved by adding a specific gene or genes to a plant, or by knocking down a gene with RNAi, to produce a desirable phenotype. The plants resulting from adding a gene are often referred to as transgenic plants. If for genetic modification genes of the species or of a crossable plant are used under control of their native promoter, then they are called cisgenic plants. Genetic modification can produce a plant with the desired trait or traits faster than classical breeding because the majority of the plant's genome is not altered. To genetically modify a plant, a genetic construct must be designed so that the gene to be added or removed will be expressed by the plant. To do this, a promoter to drive transcription and a termination sequence to stop transcription of the new gene, and the gene or genes of interest must be introduced to the plant. A marker for the selection of transformed plants is also included. In the laboratory, antibiotic resistance is a commonly used marker: Plants that have been successfully transformed will grow on media containing antibiotics; plants that have not been transformed will die. In some instances markers for selection are removed by backcrossing with the parent plant prior to commercial release.

The construct can be inserted in the plant genome by genetic recombination using the bacteria *Agrobacterium tumefaciens* or *A. rhizogenes*, or by direct methods like the gene gun or microinjection. Using plant viruses to insert genetic constructs into plants is also a possibility, but the technique is limited by the host range of the virus. For example, Cauliflower mosaic virus (CaMV) only infects cauliflower and related species. Another limitation of viral vectors is that the virus is not usually passed on the progeny, so every plant has to be inoculated.

The majority of commercially released transgenic plants are currently limited to plants that have introduced resistance to insect pests and herbicides. Insect resistance is achieved through incorporation of a gene from *Bacillus thuringiensis* (Bt) that encodes a protein that is toxic to some insects. For example, the cotton bollworm, a common cotton pest, feeds on Bt cotton it will ingest the toxin and die. Herbicides usually work by binding to certain plant enzymes and inhibiting their action. The enzymes that the herbicide inhibits are known as the herbicides *target site*. Herbicide resistance can be engineered into crops by expressing a version of *target site* protein that is not inhibited by the herbicide. This is the method used to produce glyphosate resistant crop plants. Genetic modification of plants that can produce pharmaceuticals (and industrial chemicals), sometimes called *pharmacrops*, is a rather radical new area of plant breeding.

Issues and Concerns

Modern plant breeding, whether classical or through genetic engineering, comes with issues of concern, particularly with regard to food crops. The question of whether breeding can have a negative effect on nutritional value is central in this respect. Although relatively little direct research in this area has been done, there are scientific indications that, by favouring certain aspects of a plant's development, other aspects may be retarded. A study published in the *Journal of the American College of Nutrition* in 2004, entitled *Changes in USDA Food Composition Data for 43 Garden Crops, 1950 to 1999*, compared nutritional analysis of vegetables done in 1950 and in 1999, and found substantial decreases in six of 13 nutrients measured, including 6% of protein and 38% of riboflavin. Reductions in calcium, phosphorus, iron and ascorbic acid were also found. The study, conducted at the Biochemical Institute, University of Texas at Austin, concluded in summary: *"We suggest that any real declines are generally most easily explained by changes in cultivated varieties between 1950 and 1999, in which there may be trade-offs between yield and nutrient content."*

The debate surrounding genetically modified food during the 1990s peaked in 1999 in terms of media coverage and risk perception, and continues today - for example, *"Germany has thrown its weight behind a growing European mutiny over genetically modified crops by banning the planting of a widely grown pest-resistant corn variety."*. The debate encompasses the ecological impact of genetically modified plants, the safety of genetically modified food and concepts used for safety evaluation like substantial equivalence. Such concerns are not new to plant breeding. Most countries have regulatory processes in place to help ensure that new crop varieties entering the marketplace are both safe and meet farmers' needs. Examples include variety registration, seed schemes, regulatory authorizations for GM plants, etc.

Plant breeders' rights is also a major and controversial issue. Today, production of new varieties is dominated by commercial plant breeders, who seek to protect their work and collect royalties through national and international agreements based in intellectual property rights. The range of related issues is complex. In the simplest terms, critics of the increasingly restrictive regulations argue that, through a combination of technical and economic pressures, commercial breeders are reducing biodiversity and significantly constraining individuals (such as farmers) from developing and trading seed on a regional level. Efforts to strengthen breeders' rights, for example, by lengthening periods of variety protection, are ongoing.

When new plant breeds or cultivars are bred, they must be maintained and propagated. Some plants are propagated by asexual means while others are propagated by seeds. Seed propagated cultivars require specific control over seed source and production procedures to maintain the integrity of the plant breeds results. Isolation is necessary to prevent cross contamination with related plants or the mixing of seeds after harvesting. Isolation is normally accomplished by planting distance but in certain crops, plants are enclosed in greenhouses or cages (most commonly used when producing F1 hybrids.)

Participatory Plant Breeding

The development of agricultural science, with phenomenon like the Green Revolution arising, have left millions of farmers in developing countries, most of whom operate small farms under unstable and difficult growing conditions, in a precarious situation. The adoption of new plant varieties by this group has been hampered by the constraints of poverty and the international policies promoting an industrialized model of agriculture. Their response has been the creation of a novel and promising set of research methods collectively known as participatory plant breeding. Participatory means that farmers are more involved in the breeding process and breeding goals are defined by farmers instead of international seed companies with their large-scale breeding programmes. Farmers' groups and NGOs, for example, may wish to affirm local people's rights over genetic resources, produce seeds themselves, build farmers' technical expertise, or develop new products for niche markets, like organically grown food.

Agricultural Knowledge, Science and Technology for Development

The International Assessment of Agricultural Knowledge, Science and Technology for Development (IAASTD) is an international effort initiated by the World Bank that evaluated the relevance, quality and effectiveness of agricultural knowledge, science, and technology (AKST), and the effectiveness of public and private sector policies and institutional arrangements. The project developed out of a consultative process involving 900 participants and 110 countries. The IAASTD was launched as an intergovernmental process, under the co-sponsorship of the FAO, GEF, UNDP, UNEP, UNESCO, the World Bank and WHO. The IAASTD was a three-year collaborative effort (2005–2007) that assessed AKST with respect to meeting development and sustainability goals of reducing hunger and poverty, improving nutrition, health and rural livelihoods, and facilitating social and environmental sustainability The results and conclusions of the project were reviewed and ratified during the

Intergovernmental Plenary Meeting held 7–12 April 2008, in Johannesburg, South Africa.

Governance and Management

The geographically based multi-stakeholder Bureau is composed of 30 government representatives from different regions, 22 representatives from non-governmental organizations); consumer groups, and producer groups, representatives from 8 institutions, and 2 co-chairs. The sponsoring agencies serve as ex-officio members of the Bureau. The IAASTD has a distributed Secretariat providing management and oversight, with the major component in Washington DC and others in FAO (Rome), UNEP (Nairobi), and UNESCO (Paris). The current Director is Robert T. Watson. Other members of the distributed Secretariat include staff located at the Sub-global Management Entities.

Overview and Structure

The IAASTD is composed of one Global Assessment and five Sub-global Assessments, which use the same framework: the impacts of AKST on hunger, poverty, nutrition, human health, and environmental and social sustainability in the past and the future. The Global and Sub-global assessments are peer-reviewed by governments and experts, and approved by the Panel of participating governments.

The five Sub-global Assessments:

- Central and West Asia and North Africa (CWANA) - Regional Institute: ICARDA (International Center for Agricultural Research in the Dry Areas)
- East and South Asia and the Pacific (ESAP) - Regional Institute: World Fish Center
- Latin America and the Caribbean (LAC) - Regional Institute: IICA (Inter-American Institute for Cooperation on Agriculture)
- North America and Europe (NAE)
- Sub-Saharan Africa (SSA)- Regional Institute: ACTS (African Centre for Technology Studies) complement the Global Assessment by examining geographic area-specific aspects.

The assessments are accompanied by a synthesis report which covers challenges faced by agriculture today: pros and cons of bioenergy, potential role of biotechnology, effects of climate change, effects on health, use of natural resources, small farmers & global trade, future role for traditional farming, women in agriculture, and options for action.

Final Report

On 15 April 2008, IAASTD report findings were released. The report incorporates a global assessment as well as five sub-global assessments in acknowledgement that the challenges in Africa are not identical to the challenges in Asia or Latin America. By taking a 'bottom-up' approach, the report aims to understand the needs of those most vulnerable to threats to the security of their food and livelihood. Prior plenary sessions in Johannesburg aimed to come to agreement on the key priorities for each region.

A series of published (printed and web-based), critical, in-depth Global and Sub-global Assessments of local and institutional knowledge and experiences was produced. The Assessment reports have been translated into the six official UN languages, presented, and discussed in multiple forums. The reports created 'plausible scenarios', based on past events and existing trends such as population growth, rural/urban food and poverty dynamics, loss of agricultural land, water availability and climate change effects. Based around these issues, 'What if?' questions were be formulated that allow the implications of different technological options to be explored.

Agricultural Machinery

Agricultural machinery is machinery used in the operation of an agricultural area or farm.

History

The Industrial Revolution

With the coming of the Industrial Revolution and the development of more complicated machines, farming methods took a great leap forward. Instead of harvesting cereal | grain by hand with a sharp blade, wheeled machines cut a continuous swath. Instead of threshing the grain by beating it with sticks, threshing machines separated the seeds from the heads and stalks.

Steam Power

Power for agricultural machinery was originally supplied by horses or other domesticated animals. With the invention of steam power came the portable engine, and later the traction engine, a multipurpose, mobile energy source that was the ground-crawling cousin to the steam locomotive. Agricultural steam engines took over the heavy pulling work of horses, and were also equipped with a pulley that could power stationary machines via the use of a long belt. The steam-powered machines were low-powered by today's standards but, because of their size and their low gear ratios, they could provide a large drawbar pull. Their slow speed led farmers to comment that tractors had two speeds: "slow, and darn slow."

Internal Combustion Engines

The internal combustion engine; first the petrol engine, and later diesel engines; became the main source of power for the next generation of tractors. These engines also contributed to the development of the self-propelled, combined harvester and thresher, or combine harvester (also shortened to 'combine'). Instead of cutting the grain stalks and transporting them to a stationary threshing machine, these combines cut, threshed, and separated the grain while moving continuously through the field.

Types

Combines might have taken the harvesting job away from tractors, but tractors still do the majority of work on a modern farm. They are used to pull implements—machines that till the ground, plant seed, and perform other tasks.

Tillage implements prepare the soil for planting by loosening the soil and killing weeds or competing plants. The best-known is the plow, the ancient implement that was upgraded in 1838 by John Deere. Plows are now used less frequently in the U.S. than formerly, with offset disks used instead to turn over the soil, and chisels used to gain the depth needed to retain moisture.

The most common type of seeder is called a planter, and spaces seeds out equally in long rows, which are usually two to three feet apart. Some crops are planted by drills, which put out much more seed in rows less than a foot apart, blanketing the field with crops. Transplanters automate the task of transplanting seedlings to the field. With the widespread use of plastic mulch, plastic mulch layers, transplanters, and seeders lay down long rows of plastic, and plant through them automatically.

After planting, other implements can be used to cultivate weeds from between rows, or to spread fertilizer and pesticides. Hay balers can be used to tightly package grass or alfalfa into a storable form for the winter months.

Modern irrigation relies on machinery. Engines, pumps and other specialized gear provide water quickly and in high volumes to large areas of land. Similar types of equipment can be used to deliver fertilizers and pesticides.

Besides the tractor, other vehicles have been adapted for use in farming, including trucks, airplanes, and helicopters, such as for transporting crops and making equipment mobile, to aerial spraying and livestock herd management.

New Technology and the Future

Though modern harvesters and planters will do a better job than their predecessors, the combine of today still cuts, threshes, and separates grain in essentially the same way it has always been done. However, technology is changing the way that humans operate the machines, as computer monitoring systems, GPS locators, and self-steer programmes allow the most advanced tractors and implements to be more precise and less wasteful in the use of fuel, seed, or fertilizer. In the foreseeable future, some agricultural machines will be capable of driving themselves, using GPS maps and electronic sensors to become agricultural robots. Even more esoteric are the new areas of nanotechnology and genetic engineering, where submicroscopic devices and biological processes may be used as machines to perform agricultural tasks.

CROP PRODUCTION SYSTEMS

Cropping systems vary among farms depending on the available resources and constraints; geography and climate of the farm; government policy; economic, social and political pressures; and the philosophy and culture of the farmer. Shifting cultivation (or slash and burn) is a system in which forests are burnt, releasing nutrients to support cultivation of annual and then perennial crops for a period of several years.

Then the plot is left fallow to regrow forest, and the farmer moves to a new plot, returning after many more years (10-20). This fallow period is shortened if population density grows, requiring the input of nutrients (fertilizer or manure) and some manual pest control. Annual cultivation is the next phase of intensity in which there is no fallow period. This requires even greater nutrient and pest control inputs.

Further industrialization lead to the use of monocultures, when one cultivar is planted on a large acreage. Because of the low biodiversity, nutrient use is uniform and pests tend to build up, necessitating the greater use of pesticides and fertilizers. Multiple cropping, in which several crops are grown sequentially in one year, and intercropping, when several crops are grown at the same time are other kinds of annual cropping systems known as polycultures. In tropical environments, all of these cropping systems are practiced. In subtropical and arid environments, the timing and extent of agriculture may be limited by rainfall, either not allowing multiple annual crops in a year, or requiring irrigation. In all of these environments perennial crops are grown (coffee, chocolate) and systems are practiced such as agroforestry. In temperate environments, where ecosystems were predominantly grassland or prairie, highly productive annual cropping is the dominant farming system. The last century has seen the intensification, concentration and specialization of agriculture, relying upon new technologies of agricultural chemicals (fertilizers and pesticides), mechanization, and plant breeding (hybrids and GMO's). In the past few decades, a move towards sustainability in agriculture has also developed, integrating ideas of socio-economic justice and conservation of resources and the environment within a farming system. This has led to the development of many responses to the conventional agriculture approach, including organic agriculture, urban agriculture, community supported agriculture, ecological or biological agriculture, integrated farming and holistic management, as well as an increased trend towards agricultural diversification.

LIVESTOCK PRODUCTION SYSTEMS

Animals, including horses, mules, oxen, camels, llamas, alpacas, and dogs, are often used to help cultivate fields, harvest crops, wrangle other animals, and transport farm products to buyers. Animal husbandry not only refers to the breeding and raising of animals for meat or to harvest animal products (like milk, eggs, or wool) on a continual basis, but also to the breeding and care of species for work and companionship. Livestock production systems can be defined based on feed source, as grassland - based, mixed, and landless. Grassland based livestock production relies upon plant material such as shrubland, rangeland, and pastures for feeding ruminant animals. Outside nutrient inputs may be used, however manure is returned directly to the grassland as a major nutrient source. This system is particularly important in areas where crop production is not feasible because of climate or soil, representing 30-40 million pastoralists. Mixed production systems use grassland, fodder crops and grain feed crops as feed for ruminant and monogastic (one stomach; mainly chickens and pigs) livestock. Manure is typically recycled in mixed systems as a fertilizer for crops. Approximately 68% of all agricultural land is permanent pastures used in the production of livestock. Landless systems rely upon feed from outside the farm, representing the de-linking of crop and livestock production found more prevalently in OECD member countries. In the U.S., 70% of the grain grown is fed to animals on feedlots. Synthetic fertilizers are more heavily relied upon for crop production and manure utilization becomes a challenge as well as a source for pollution.

Production Practices

Tillage is the practice of plowing soil to prepare for planting or for nutrient incorporation or for pest control. Tillage varies in intensity from conventional to no-till. It may improve productivity by warming the soil, incorporating fertilizer and controlling weeds, but also renders soil more prone to erosion, triggers the decomposition of organic matter releasing CO_2, and reduces the abundance and diversity of soil organisms.

Pest control includes the management of weeds, insects/mites, and diseases. Chemical (pesticides), biological (biocontrol), mechanical (tillage), and cultural practices are used. Cultural practices include crop rotation, culling, cover crops, intercropping, composting, avoidance, and resistance. Integrated pest management attempts to use all of these methods to keep pest populations below the number which would cause economic loss, and recommends pesticides as a last resort.

Nutrient management includes both the source of nutrient inputs for crop and livestock production, and the method of utilization of manure produced by livestock. Nutrient inputs can be chemical inorganic fertilizers, manure, green manure, compost and mined minerals. Crop nutrient use may also be managed using cultural techniques such as crop rotation or a fallow period. Manure is used either by holding livestock where the feed crop is growing, such as in managed intensive rotational grazing, or by spreading either dry or liquid formulations of manure on cropland or pastures. Water management is where rainfall is insufficient or variable, which occurs to some degree in most regions of the world. Some farmers use irrigation to supplement rainfall. In other areas such as the Great Plains in the U.S. and Canada, farmers use a fallow year to conserve soil moisture to use for growing a crop in the following year. Agriculture represents 70% of freshwater use worldwide.

Processing, Distribution, and Marketing

In the United States, food costs attributed to processing, distribution, and marketing have risen while the costs attributed to farming have declined. This is related to the greater efficiency of farming, combined with the increased level of value addition (e.g. more highly processed products) provided by the supply chain. From 1960 to 1980 the farm share was around 40%, but by 1990 it had declined to 30% and by 1998, 22.2%. Market concentration has increased in the sector as well, with the top 20 food manufacturers accounting for half the food-processing value in 1995, over double that produced in 1954. As of 2000 the top six US supermarket groups had 50% of sales compared to 32% in 1992. Although the total effect of the increased market concentration is likely increased efficiency, the changes redistribute economic surplus from producers (farmers) and consumers, and may have negative implications for rural communities.

Crop Alteration and Biotechnology

Crop alteration has been practiced by humankind for thousands of years, since the beginning of civilization. Altering crops through breeding practices changes the genetic make-up of a plant to develop crops with more beneficial characteristics for humans, for example, larger fruits or seeds, drought-tolerance, or resistance to pests. Significant advances in plant breeding ensued after the work of geneticist Gregor Mendel. His work on dominant and recessive alleles gave plant breeders a better understanding of genetics and brought great insights to the techniques utilized by plant breeders. Crop breeding includes techniques such as plant selection with desirable traits, self-pollination and cross-pollination, and molecular techniques that genetically modify the organism. Domestication of plants has, over the centuries increased yield, improved disease resistance and drought tolerance, eased harvest and improved the taste and nutritional value of crop plants. Careful selection and breeding have had enormous effects on the characteristics of crop plants. Plant selection and breeding in the 1920s and 1930s improved pasture (grasses and clover) in New Zealand. Extensive X-ray an ultraviolet induced mutagenesis efforts (i.e. primitive genetic engineering) during the 1950s produced the modern commercial varieties of grains such as wheat, corn (maize) and barley.

The Green Revolution popularized the use of conventional hybridization to increase yield many folds by creating "high-yielding varieties". For example, average yields of corn (maize) in the USA have increased from around 2.5 tons per hectare (t/ha) (40 bushels per acre) in 1900 to about 9.4 t/ha (150 bushels per acre) in 2001. Similarly, worldwide average wheat yields have increased from less than 1 t/ha in 1900 to more than 2.5 t/ha in 1990. South American average wheat yields are around 2 t/ha, African under 1 t/ha, Egypt and Arabia up to 3.5 to 4 t/ha with irrigation. In contrast, the average wheat yield in countries such as France is over 8 t/ha. Variations in yields are due mainly to variation in climate, genetics, and the level of intensive farming techniques (use of fertilizers, chemical pest control, growth control to avoid lodging).

Genetic Engineering

Genetically Modified Organisms (GMO) are organisms whose genetic material has been altered by genetic engineering techniques generally known as recombinant DNA technology. Genetic engineering has expanded the genes available to breeders to utilize in creating desired germlines for new crops. After mechanical tomato-harvesters were developed in the early 1960s, agricultural scientists genetically modified tomatoes to be more resistant to mechanical handling. More recently, genetic engineering is being employed in various parts of the world, to create crops with other beneficial traits. New research on woodland strawberry genome was found to be short and easy to manipulate. Researchers now have tools to improve strawberry flavors and aromas of cultivated strawberries as stated in a publication by Nature Genetics. http://www.isaaa.org/kc/cropbiotechupdate/article/default.asp?ID=7160

Herbicide-Tolerant GMO Crops

Roundup Ready seed has a herbicide resistant gene implanted into its genome that allows the plants to tolerate exposure to glyphosate. *Roundup* is a trade name for a glyphosate-based product, which is a systemic, nonselective herbicide used to kill weeds. *Roundup Ready* seeds allow the farmer to grow a crop that can be sprayed with glyphosate to control weeds without harming the resistant crop. Herbicide-tolerant crops are used by farmers worldwide. Today, 92% of soybean acreage in the US is planted with genetically modified herbicide-tolerant plants.

With the increasing use of herbicide-tolerant crops, comes an increase in the use of glyphosate-based herbicide sprays. In some areas glyphosate resistant weeds have developed, causing farmers to switch to other herbicides. Some studies also link widespread glyphosate usage to iron deficiencies in some crops, which is both a crop production and a nutritional quality concern, with potential economic and health implications.

Insect-resistant GMO Crops

Other GMO crops used by growers include insect-resistant crops, which have a gene from the soil bacterium *Bacillus thuringiensis* (Bt), which produces a toxin specific to insects. These crops protect plants from damage by insects; one such crop is Starlink. Another is cotton, which accounts for 63% of US cotton acreage.

Some believe that similar or better pest-resistance traits can be acquired through traditional breeding practices, and resistance to various pests can be gained through hybridization or cross-pollination with wild species. In some cases, wild species are the primary source of resistance traits; some tomato cultivars that have gained resistance to at least 19 diseases did so through crossing with wild populations of tomatoes.

Costs and Benefits of GMOs

Genetic engineers may someday develop transgenic plants which would allow for irrigation, drainage, conservation, sanitary engineering, and maintaining or increasing yields while requiring fewer fossil fuel derived inputs than conventional crops. Such developments would be particularly important in areas which are normally arid and rely upon constant irrigation, and on large scale farms. However, genetic engineering of plants has proven to be controversial. Many issues surrounding food security and environmental impacts have risen regarding GMO practices. For example, GMOs are questioned by some ecologists and economists concerned with GMO practices such as terminator seeds, which is a genetic modification that creates sterile seeds. Terminator seeds are currently under strong international opposition and face continual efforts of global bans.

Another controversial issue is the patent protection given to companies that develop new types of seed using genetic engineering. Since companies have intellectual ownership of their seeds, they have the power to dictate terms and conditions of their patented product. Currently, ten seed companies control over two-thirds of the global seed sales. Vandana Shiva argues that these companies are guilty of biopiracy by patenting life and exploiting organisms for profit Farmers using patented seed are restricted from saving seed for subsequent plantings, which forces farmers to buy new seed every year. Since seed saving is a traditional practice for many farmers in both developing and developed countries, GMO seeds legally bind farmers to change their seed saving practices to buying new seed every year.

Locally adapted seeds are an essential heritage that has the potential to be lost with current hybridized crops and GMOs. Locally adapted seeds, also called land races or crop eco-types, are important because they have adapted over time to the specific microclimates, soils, other environmental conditions, field designs, and ethnic preference indigenous to the exact area of cultivation. Introducing GMOs and hybridized commercial seed to an area brings the risk of cross-pollination with local land races Therefore, GMOs pose a threat to the sustainability of land races and the ethnic heritage of cultures. Once seed contains transgenic material, it becomes subject to the conditions of the seed company that owns the patent of the transgenic material.

Modern Agriculture

Modern agriculture is a term used to describe the wide majority of production practices employed by America's farmers. The term depicts the push for innovation, stewardship and advancements continually made by growers to sustainably produce higher-quality products with a reduced environmental impact. Intensive scientific research and robust investment in modern agriculture during the past 50 years has helped farmers double food production while essentially freezing the footprint of total cultivated farmland.

Safety

The agriculture industry works with government agencies and other organizations to ensure that farmers have access to the technologies required to support modern agriculture practices. Farmers are supported by education and certification programmes that ensure they apply agricultural practices with care and only when required.

Sustainability

Technological advancements help provide farmers with tools and resources to help reduce their environmental footprint and to make farming more sustainable. New technologies have given rise to innovations like conservation tillage, a farming process which helps prevent land loss to erosion, water pollution and enhances carbon sequestration. The World Bank, the Bill & Melinda Gates Foundation and others have noted that integrated crop management is based on agro-ecological principles and can increase yields while reducing environmental damage.

Affordability

The goal of modern agriculture practices is to help farmers provide an affordable supply of food to meet the demands of a growing population. With modern agriculture, more crops can be grown on less land allowing farmers to provide an increased supply of food at an affordable price.

9

Macro Management of Agriculture

Agriculture being a State subject, the primary responsibility for increasing agriculture production, enhancing productivity and exploring the vast untapped potential of the sector rests with State Governments. However, in order to supplement the efforts of State Governments, a number of Centrally Sponsored and Central Sector Schemes are being implemented as a catalyst for enhancement of agricultural production and productivity in the country, and consequential increase in prosperity of the farming community.

The Macro Management of Agriculture (MMA) Scheme is a Centrally Sponsored Scheme formulated with the objective to ensure that the Central Assistance is spent on focused and specific interventions for development of agriculture in areas of priority of different States. It became operational in 2000-01 in all States and UTs. The Scheme provides sufficient flexibility to States to develop and pursue the programmes on the basis of their regional priorities. Thus States have been given a free hand to finalise their sector-wise allocation as per requirements of their developmental priorities. It is a major step towards achieving decentralisation in pursuance of restoring primacy of States in agricultural development planning which aims at development in agriculture through Work Plans prepared by State Governments themselves.

The scheme initially consisted of 27 Centrally Sponsored Schemes relating to Cooperative, Crop Production Programmes (for rice, wheat, coarse cereals, jute, sugarcane) Watershed Development Programme (NWDPRA, RVP/FPR), Horticulture, Fertiliser, Mechanisation and Seeds Production Programme. However, with the launching of National Horticulture Mission (NHM), 10 schemes pertaining to horticulture development were taken out of the purview of this Scheme from the year 2005-06.

The approved pattern of Assistance under the Scheme is in the ratio of 90:10 between the Centre and the State respectively except in the case of North Eastern States in case of which 100 per cent Central Assistance is extended. The Central Assistance to the States is released in two installments in the ratio of 80 per cent Grant and 20 per cent Loan.

The position regarding Budget Estimates, Revised Estimates and expenditure incurred under the scheme since 2000-01 is as indicated below (Rs in Crore):

Year	Budget Estimate	Revised Estimate	Expenditure
2000-01	490.00	381.88	381.88
2001-02	850.00	680.49	678.62
2002-03	736.86	597.00	597.59
2003-04	700.00	648.60	648.49
2004-05	712.92	1189.20	1188.94
2005-06	912.62	819.15	841.86

An amount of Rs 910.00 crore has been earmarked in BE 2006-07 for implementation of this Scheme.

AGRICULTURAL MARKETING

Organised marketing of agricultural commodities has been promoted in the country through a network of regulated markets. Most of the State governments and Union Territories have enacted legislations (APMC Act) to provide for regulation of agricultural produce markets. While by the end of 1950, there were 286 regulated markets in the country, today the number stands at 7,521 (31.3.2005).

Besides, the country has 27,294 rural periodical markets, about 15 per cent of which function under the ambit of regulation. The advent of regulated markets has helped in mitigating the market handicaps of producers/sellers at the wholesale assembling level. But, the rural periodic markets in general, and the tribal markets in particular, remained out of its developmental ambit.

Agriculture sector needs well functioning markets to drive growth, employment and economic prosperity in rural areas of the country. In order to provide dynamism and efficiency into the marketing system, large investments are required for the development of post harvest and cold chain infrastructure nearer to the farmers' field. A major portion of this investment is expected from the private sector, for which an appropriate regulatory and policy environment is necessary.

Alongside, enabling policies need to be put in place to encourage procurement of agricultural commodities directly from farmers' field and to establish effective linkage between the farm production and the retail chain and food processing industries. Accordingly, amendment to the State APMC Act for deregulation of marketing system in the country is suggested to promote investment in marketing infrastructure, motivating corporate sector to undertake direct marketing and to facilitate a national integrated market.

The Ministry of Agriculture formulated a model law on agricultural marketing for guidance and adoption by State Governments. The model legislation provides for establishment of Private Markets/Yards, Direct Purchase Centres, Consumer/Farmers Markets for direct sale and promotion of Public Private Partnership in the management and development of agricultural markets in the country. Provision has also been made in the Act for constitution of State Agricultural Produce Marketing Standards Bureau for promotion of Grading, Standardisation and Quality Certification of agricultural produce. This would facilitate pledge financing, direct purchasing, forward/future trading and exports. Several States have initiated steps for amending the APMC Act.

- *Infrastructure Requirement:* Investment requirement for the development of marketing, storage and cold storage infrastructure in the country has been estimated to be huge and with a view to induce investment in the development of marketing infrastructure as envisaged above, the Ministry has implemented the following Plan Schemes:
  i. A capital investment subsidy scheme titled "Construction of Rural Godowns" is being implemented w.e.f. 1 April 2001. The main objectives of the scheme include creation of scientific storage capacity with allied facilities in rural areas to meet various requirements of farmers for storing farm produce, processed farm produce, agricultural inputs, etc., and prevention of distress sale by creating the facility of pledge loan and marketing credit. Under the original scheme, back ended subsidy @ 25 per cent of capital cost of the project was provided. In case of NE States, hilly areas and SC/ST entrepreneurs, subsidy was provided @ 33.33 per cent of the capital cost of the project. The Scheme has since been modified with effect from 20 October 2004, to provide subsidy @ 25 per cent to farmers, Agriculture graduates, cooperatives and Central Warehousing Corporation/State Warehousing Corporations. All other categories of individuals companies and corporations are now given subsidy @ 15 per cent of the project cost. The scheme has been made farmers' friendly by allowing subsidy for smaller

godowns of 50 MT size in general and of 25 in hilly areas. Five lakh tonnes capacity to be created is reserved for small farmers. The scheme is being implemented through NABARD and NCDC. Till 31 May 2006, 11,583 storage projects having a capacity of 166.42 lakh tonnes have been sanctioned under the scheme.

ii. With a view to establish a nation-wide information network for speedy collection and dissemination of price and market related information to farmers, electronic connectivity is being provided to all important agricultural markets in the country under a Central scheme, "Market Research and Information Network". 2,408 market nodes and 92 State Marketing Boards and Directorate of Marketing and Inspection offices have been networked on a single portal, wherein daily prices of more than 300 commodities and about 2000 varieties are being reported. It is planned to connect 2,700 markets in all, under the scheme during the 10th Plan.

iii. The Ministry of Agriculture is implementing another Central Sector scheme for "Development/Strengthening of Agricultural Marketing Infrastructure, Grading and Standardisation". Under the scheme, investment subsidy is provided @ 25 per cent on the capital cost of the marketing infrastructure development project subject to a maximum of Rs 50 lakh for each project in all States and @ 33.3 per cent of capital cost subject to a maximum of Rs 60 lakh for each project in case of North Eastern States, hilly areas and to Scheduled Castes/Scheduled Tribes entrepreneurs. In respect of infrastructure projects of State Governments/State Agencies, there is no upper ceiling on subsidy to be provided under the scheme. The scheme is reform linked, to be implemented in those States/Union Territories wherein the law dealing with agriculture markets (Agricultural Produce Marketing Regulation Act) allows setting up of competitive agricultural markets in private and cooperative sectors, direct marketing and contract farming. The States of Andhra Pradesh, Punjab, Kerala, Tamil Nadu, Manipur, Sikkim, Madhya Pradesh, Himachal Pradesh, Nagaland, Rajasthan, Chattisgarh and Union Terriroty of Andaman and Nicobar Islands, Daman and Diu and Dadra and Nagar Haveli have notified to receive assistance under the Scheme. The remaining States/UTs are in the process of amending their APMC Acts. 158 training and awareness programmes have been conducted in the notified States/UTs. A total number of 259 new project proposals have been provided advance subsidy of Rs 516.30 lakh by NABARD in the States of Madhya Pradesh, Tamilnadu, Punjab, Andhra Pradesh and Kerala.

iv. The Department has recently taken the initiative to promote modern terminal markets for fruits, vegetables and other perishables in important urban centres of the country. These markets would provide state of art infrastructure facilities for electronic auction, cold chain and logistics and operate through primary collection centres conveniently located in producing areas to allow easy access to farmers. The terminal markets are envisaged to operate on a 'Hub-and-Spoke' format wherein the Terminal Market (the hub) would be linked to a number of collection centres (the spokes), conveniently located in key production centers to allow easy access to farmers for the marketing of their produce. The concept on setting up of Terminal Markets for perishable commodities was discussed with the State Governments and interested private enterprises at a national conference of State Ministers held on 20 February 2006 at New Delhi. Based on the discussions, a Committee has been constituted under the Chairmanship of Director General, National Institute of Agricultural Marketing, Jaipur with members from participating State Governments to develop a framework for the bidding process for selecting the enterprise for the implementation of terminal market projects and to work out implementation modalities. Central assistance to these projects is planned by way of equity participation.

The Department of Agriculture and Cooperation has three organisations dealing with marketing under its administrative control, namely, the Directorate of Marketing and Inspection (DMI), Faridabad, the Ch. Charan Singh National Institute of Agricultural Marketing (NIAM), Jaipur and the Small Farmers Agri-Business Consortium (SFAC), New Delhi.

- *Directorate of Marketing and Inspection:* It is an attached office of the Department and is headed by Agricultural Marketing Adviser. The Directorate has its Head Office at Faridabad (Haryana), Branch Head Office at Nagpur (Maharashtra), 11 Regional Offices and the Central Agmark Laboratory at Nagpur. Besides, there are 26 Sub-Offices, 16 Regional Agmark Laboratories (RALs) spread all over the country.

  The main functions of the Directorate are as follow:

  1. Rendering advice on statutory regulation, development and management of agricultural produce markets to the States/UTs
  2. Promotion of Standardisation and Grading of agricultural and allied produce under the Agricultural Produce (Grading and Marking) Act, 1937

3. Market Research, Surveys and Planning
4. Training of personnel in Agricultural Marketing
5. Marketing Extension
6. Agricultural Marketing Information Network
7. Construction of Rural Godowns and
8. Development of Agricultural Marketing Infrastructure.

- *Grading and Standardisation:* The Agricultural Produce (Grading and Marking) Act, 1937 empowers the Government to fix quality standards, known as "AGMARK" standards and to prescribe terms and conditions for using the seal of 'AGMARK'. So far, grade standards have been notified for 182 agricultural and allied commodities. The purity standards under the provision of the Prevention of Food Adulteration (PFA) Act, 1954 and Bureau of Indian Standards (BIS) Act, 1986 are invariably taken into consideration while framing the grade standards. International Standards framed by Codex/International Standards Organisation (ISO) are also considered so that Indian produce can compete in the international market.

  During the year 2005-06, the final notification of Spices G & M Rules, 2005 is published, containing standards of eleven spices, namely, large Cardamom, Cardamom, Turmeric, Chillies, Ginger, Black Pepper, Coriander, Fennel, Fenugreek, Celery, Cumin, etc. The standards of Walnut duly harmonised with the international standards have been submitted to APEDA and were discussed in the Core Group's Meeting. The approval of Standing Committee on fresh Fruits and Vegetables is awaited for commodities like Lemons, Limes, Mandarins, Oranges, Grape Fruits, Walnuts in-shell and Walnut shelled.

- *National Institute of Agricultural Marketing:* The National Institute of Agricultural Marketing (NIAM) started functioning at Jaipur (Rajasthan) from 8 August 1988. NIAM has been imparting training to senior and middle level executives of agricultural and horticultural departments, Agro Industries, Corporations, State Marketing Boards, Agricultural Produce Market Committees and Apex level Cooperatives, Commodity Boards, export houses recognised by Agricultural and Processed Food Products Export Development Agency (APEDA), Commercial Banks and non-governmental organisations. Besides these clients, the NIAM also imparts training to farmers on marketing management. The main objectives of NIAM are:

  0. To provide specialised training in agricultural marketing designed to develop leadership potential in the management of agricultural marketing enterprises and services

1. To undertake research in agricultural marketing for Government, Cooperative and other Institutes, both on public funding and by contract
2. To undertake appraisal of markets/marketing projects for approval and financial support by the Central Government, on consultancy basis
3. To formulate objective criteria for selective development of physical markets and to evolve a practical methodology for the application of such criteria in their planning
4. To offer advisory and consultant services on marketing policies, investment programmes and marketing development strategies and specific advice to marketing enterprises (State, Private and Cooperatives)
5. To survey, study and analyse the rural market management and to examine in depth the principal and practice of market regulation as a development sector in the agricultural economy.

   The NIAM is managed by a Governing Body under the Chairmanship of Minister of Agriculture and an Executive Committee under the Chairmanship of Secretary, Department of Agriculture and Cooperation.

- *Training Activities:* The Institute organises Training Programmes for officials, farmers and other functionaries. In 2003-04 Management Development Programmes (MDP) were introduced. So far four MDPs have been successfully organised for leading companies like Bayer Crop Science, MICO BOSCH, etc. The MDPs were held at NIAM campus and 120 working executives of these MNCs working in various capacities have been benefited from this programme. All these MDPs were rated excellent by the participants and many other companies have shown keen interest in organising MDPs at NIAM, Jaipur.
- *Research:* The domestic Agricultural Marketing scenario has witnessed lot of changes in policies and regulations. The enactment of Model Act by some States have brought substantial improvements in trade and marketing. Following Research studies are being conducted by NIAM during the year 2005-06.

  0. Contract Farming-prospects and implications
  1. Commodity Trade Research
  2. Market led extension-a participatory approach
  3. Developing India GAP Standards

4. Information need assessment of stakeholders in Agricultural Marketing-A case of Rajasthan

- *Project Formulation:* In order to generate resources and ensure optimum utilisation of the expertise of the NIAM faculty, the Institute is taking up several Consultancy Projects in the year 2005-06. These include setting up Modern Terminal Market for fruits and vegetables at Nasik, Chandigarh, Nagpur, Patna, Bhopal, Rai (Haryana), Multi-utility integrated facility Centre-Pack house at Ludhiana for MARKFED, Punjab, State Master Plan for Market Development in Orissa, Price Forecasting for Agricultural Commodity in Karnataka, On-line Market information system for Karnataka, Designing, Planning and detailed Project report for CA Storage at Kolkata. Detailed Project Reports have been prepared for the Modern Terminal Markets at Nashik, Nagpur, Chandigarh, Bhopal and Rai (Haryana).
- *Post Garduate Programme in Agri-Business Managment (PGPABM):* The Institute has undertaken Post-Graduate Programme in Agri-Business Management (PGPABM) as a Sub-centre of MANAGE, Hyderabad from July 2001 and 50 students are presently undergoing Post-Graduate Programme in Agri-Business. The programme is designed to assist agricultural graduates to acquire the critical know-how to compete in the domestic and global business arena and to make them efficient agri-business managers.
- *Small Farmers Agri-Business Consortium:* The Small Farmers Agri-business Consortium (SFAC) was registered by Department of Agriculture and Cooperation as a Society under the Societies Registration Act, 1860 on 18 January 1994. Members at present include RBI, SBI, IDBI, EXIM Bank, Oriental Bank of Commerce, NABARD, Canara Bank, NAFED, United Phosphorous Ltd., etc.

  The SAFC is managed by a Board of Managers consisting of 20 members and chaired by Hon'ble Union Minister of Agriculture as its Ex-Officio President and the Secretary (Department of Agriculture and Cooperation), Government of India as its Ex-Officio Vice President. Managing Director is the Chief Executive of SFAC. SFAC has established 18 State level SFACs by contributing a corpus fund. The mission of the Society is to support innovative ideas for generating income and employment in rural areas by promoting private investments in agribusiness projects.

  The Central sector scheme for agri-business development implemented by SFAC was approved by the Government on 19 July 2005 for implementation during remaining period of the Tenth Plan with an outlay of Rs 48 crore. The scheme is being implemented by SFAC in close association with commercial banks for providing

- Venture Capital Assistance to agribusiness projects and assist farmer/ producer groups in preparation of quality Detailed Project Reports (DPR).

The main objectives of the scheme are to facilitate setting up of agribusiness ventures in participation with banks, catalyse private investment in setting up of agribusiness projects and thereby providing assured market to producers for increasing rural income and employment, strengthen backward linkages of agri-business projects with producers, assist farmers, producer groups, and agriculture graduates to enhance their participation in value chain through project development facility, arrange training and visits, etc., of agripreneurs setting up identified agribusiness projects. SFAC provides financial assistance to agribusiness projects by way of equity participation.

The quantum of SFAC venture capital assistance depended on the project cost and will be the lowest of the following

- 10 per cent of the total project cost assessed by the bank
- 26 per cent of the project equity
- Rs 75 lakh

Higher venture capital assistance can be considered by SFAC to deserving projects on merit and/or to projects that are located in remote and backward areas, North-eastern and hilly States and projects recommended by State agencies.

The outlay for implementation of the scheme during 2005-2006 was Rs. 10 crore which has been utilised. During the year venture capital assistance has been sanctioned to 44 agribusiness projects and assistance has been provided for preparation of 11 Detailed Project Reports(DPRs).

The allocation for implementation of the scheme during 2006-07 is Rs. 38 crore.

COOPERATION

- *Cooperative Reforms:* The Cooperative movement in India traces its origin to the agriculture and allied sector and was evolved as a mechanism for pooling meagre resources of the people with a view to providing them the advantages of economies of scale. After independence, the cooperatives were considered as part of the strategy for planned economic development. The cooperatives are today at the crossroads of existence particularly in view of the fast emerging scenario of economic liberalisation and globalisation. These institutions in general suffer from resource constraint, poor governance and management, inefficiency and unviability. The cooperative reforms are, therefore, absolutely vital to the future of the cooperatives.

- *National Policy on Cooperatives:* The Union Government has formulated a National Policy on Cooperatives in consultation with States. The objective of the National Policy is to facilitate all-round development of the cooperatives in the country and to work as guiding force for the States towards successful cooperatives. The policy, envisages that cooperatives be provided necessary support, encouragement and assistance and to ensure that they work as autonomous self-reliant and democratically managed institutions accountable to their members.
- *Multi-State Cooperative Societies (MSCS):* The Central Government has taken a lead in providing required autonomy to the cooperatives by enacting MSCS Act, 2002, replacing the MSCS Act, 1984. It aims at providing functional autonomy and democracy applicable to National level cooperative societies/federations and other Multi-State Cooperative societies. It is expected that it will work as a Model Act for reforms in State Cooperative Laws.
- *The NCDC (Amendment) Act, 2002:* The NCDC Act, 1962 has been amended vide NCDC (Amendment) Act, 2002 expanding the mandate of NCDC to include food stuff, industrial goods, livestock and services in the programmes and activities in addition to the existing programmes. The definition of agriculture produce has been amended to cover wider range of activities by including edible and non-edible oilseeds cattle feed, produce of horticulture and animal husbandry, forestry, poultry, farming, pisiculture and other allied activities to agriculture. Vide the amended Act, definition of industrial goods and livestock have been elaborated covering the products of allied industries in the rural areas and includes any handicrafts or rural crafts. The livestock includes all animals to be raised for milk, meat, fleece, skin, wool and other by-products. NCDC will be able to provide loan directly to the cooperative without State/Central Government guarantee on furnishing of security to the satisfaction of the NCDC. So far activities concerning water conservation, animal care/health, disease prevention, agricultural insurance an agricultural credit, rural sanitation/drainage/sewage have been made as notified services.
- *Amendment to the Constitution in Respect of Cooperatives:* In spite of the felt need for amendment in the State Cooperative Societies Act, the pace of reforms in Cooperative legislations by the States is not encouraging.

Therefore, to ensure the democratic, autonomous and professional functioning of cooperatives, it has been decided to initiate a proposal for amendment to the Constitution for the purpose. This decision has been taken after careful deliberations with the States in a conference of State Cooperative Ministers held on 7 December 2004. The proposed amendment to the Constitution is purported to address the key issues for empowerment of cooperatives through their voluntary formation, autonomous functioning, democratic control and professional management. The Constitution One Hundred and Sixth Amendment Bill 2006 has been introduced in the Lok Sabha on 22 May 2006.

- *Constitution of High Powered Committee:* To review the achievements of Cooperative movement during the last 100 years and challenges before it, and to suggest ways and means to face them and to give a new direction to the movement, a high-powered committee has been constituted with following Terms of Reference:
    1. To review the achievements of the cooperatives during the last 100 years.
    2. To identify the challenges being faced by the cooperative sector and to suggest measures to address them to enable the movement to keep pace with the changing socio-economic environment.
    3. To suggest an appropriate policy and legislative framework and changes required in the cooperative legislation in the country with a view to ensure the democratic, autonomous and professional functioning of cooperatives.
- *Revamping of Cooperative Credit Structure:* In August 2004, Government of India constituted a Task Force under the Chairmanship of Prof. A Vaidyanathan for suggesting measures for revival of cooperative credit institutions. The Task Force has submitted its Report in respect of short-term cooperative credit structure and recommended a financial package of Rs 14,839 crore for the short-term rural credit cooperative institutions. The package covers accumulated losses, unpaid invoke guarantees, receivables from State Governments, return of share capital to State Governments, Human Resources Development, conduct of special audit, computerisation, implementation costs, etc.

Based on the consensus arrived at with State Governments and other stakeholders on the recommendations made by the Task Force, the Government has approved the package for revival of the Short-term Rural Cooperative Credit Structure involving financial assistance of Rs. 13,596 crore.

The provision of financial assistance under the package has been linked to reforms in the cooperative sector. The same Task Force has been assigned the mandate for suggesting revival package for the Long-Term Cooperative Credit Structure. The Government will initiate steps for reviving Long-Term Cooperative Credit Structure after reviving the report of the Task Fore in this regard.

ANIMAL HUSBANDRY

Animal Husbandry and Dairy Development plays a prominent role in the rural economy in supplementing the income of rural households, particularly, the landless and small and marginal farmers.

It also provides subsidiary occupation in semi-urban areas and more so for people living in hilly, tribal and drought prone areas where crop output may not sustain the family. Animal husbandry output constitutes about 30 per cent of the country's agricultural output. According to Central Statistical Organisation (CSO), the value of output from livestock and fisheries sectors is about 37.7 per cent of total agriculture and allied sectors.

India is endowed with the largest livestock population in the world. It accounts for 57 per cent of the world's buffalo population and 14 per cent of the cattle population. According to Livestock Census (2003), the country has about 18.5 crore cattle and 9.8 crore buffaloes.

- *Contribution of Livestock Sector to Food Basket:* The contribution of livestock sector to the food basket in the form of milk, eggs, and meat has been immense in fulfilling the animal protein requirement of ever-growing human population. The present availability of human protein in an Indian diet is 10 gm per person per day, as against a world average of 25 gm. However, keeping in view the growing population, the animal protein availability has to increase at least two-fold for maintaining the nutritional level of growing children and nursing mothers in India.
  - *Milk Production:* During the past five-year plans, several measures have been initiated by the Government to increase the productivity of livestock, which has resulted in significant increase in milk production to the level of 91 million tonnes in the year 2004-05. India has become the largest producer of milk in the world.
  - *Egg Production:* The wool production in the country was 44.56 million kg during 2004-05.
  - *Other Livestock Products:* Livestock sector not only provides essential protein and nutritious human diet through milk, eggs, meat, etc., but also plays an important role in utilisation of non-edible agricultural by-products. Livestock also provides raw material/by-products such as hides and skins, blood, bone, fat, etc.

* *Livestock Insurance:* The Cabinet Committee on Economic Affairs has approved the 'Livestock Insurance Scheme' on 21 February 2006 for its implementation during 2005-06 and 200-07 of the Tenth Plan on pilot basis in 100 select districts across the country. The scheme has been formulated with the twin objective of providing protection mechanism to the farmers and cattle rearers against any eventual loss of their animals and to demonstrate the benefit of the scheme to the people and popularise it with the ultimate goal of attaining qualitative improvement in livestock and their products. The scheme will cover crossbred and high yielding cattle and buffaloes during the pilot stage. The scheme envisages 50 per cent subsidy of the insurance premium. The entire expenditure on subsidised insurance premium, payment of honorarium to the veterinary practitioners and funds for publicity of the scheme will be borne by the Central Government.

  The Insurance subsidy in the pilot scheme will be provided for insurance of an amount for a maximum period of three years and will be limited to two animals per beneficiary. No subsidy on insurance will be provided for the second time/extension of insurance of same animal. The scheme is to be implemented through the State Implementing Agencies (SIA) implementing the National Project on Cattle and Buffalo Breeding Programme (NPCBB) and through State Animal Husbandry Departments in those states where SIAs do not exist.

- *Cattle And Buffalo Development:* India possesses 27 acknowledged indigenous breeds of cattle and seven breeds of buffaloes. Various Central and Centrally sponsored schemes are being implemented for genetic improvement of cattle and buffalo with a view to enhance the per capita availability of consumption of milk through increased milk production. Efforts are also made to protect and preserve the indigenous cattle and buffalo in their native tract, which are facing threat of extinction. The elite animals are selected and registered on the basis of their performance for production of superior pedigree bulls, bull-mothers, frozen semen and frozen embryos for future breeding improvements.

  The National Project for Cattle and Buffalo Breeding envisages 100 per cent grant-in-aid to implementing agencies. The Project will also promote about 14,000 private artificial insemination (AI) practitioners and buildup an annual frozen semen production capacity of 66 million doses. Since inception, 26 States have been assisted with Rs 202.52 crore upto 31 March 2005 for participating in the Project.

A Central Herd Registration Scheme for identification and location of superior germ plasm of cattle and buffaloes, propagation of superior germ stock, regulating the sale and purchase, help in formation of breeders' society and to meet requirement of superior bulls in different parts of the country is also being implemented. The Government has established Central Herd Registration Unit in four breeding tracts, i.e., Rohtak, Ahmedabad, Ongole, Ajmer. A total of 92 milk-recording centres are functioning to register these breeds of cattle, viz., Gir, Kankrej, Hariana and Ongole and in buffalo Jaffrabadi, Mehsani, Murrah and Surti. During 2005-06, 13549 animals were primary registered.

The seven Central Cattle Breeding Farms at Suratgarh (Rajasthan), Chiplima and Sunabeda (Orissa), Dhamrod (Gujarat), Hessarghatta (Karnataka), Alamadi (Tamil Nadu) and Andeshnagar (Uttar Pradesh) are engaged in scientific breeding programmes of cattle and buffaloes and production of high pedigreed bulls for National Project for Cattle and Buffalo Breeding Programme besides providing training to farmers and breeders. During 2004-05, these farms produced 313 bull calves and supplied 252 high pedigreed bull calves for use under Artificial Insemination Programme in various parts of the country. The Central Frozen Semen Production and Training Institute, Hessarghatta (Karnataka) produced 13.43 lakh doses and supplied 14.74 lakh doses of frozen semen of high pedigreed Sahiwal, Red Singhi, Holstein Friesian, Jersey, crossbred and Murrah buffalo to different States for their AI programmes.

- *Poultry Development:* The value of output from poultry sector is nearly 20000 crore and it provides direct or indirect employment to over 2 million people. About 25 per cent of the total egg production in the country comes from desi poultry, which is unorganised rural backyard system. A target for achieving production of over 52 billion eggs by 2011-12, at a growth rate of 4.3 per cent has been visualised by the Government of India.

Poultry sector, besides employment generation and subsidiary income increase, provides nutritional security especially to the rural poor. Further, landless labourers derive more than 50 per cent of their income from livestock especially poultry.

Central Poultry Development Organisations have been playing a pivotal role in the implementation of the policies of the Government with respect to poultry as a tool for alleviating nutritional hunger and palliating the impecuniosities of the resource-poor farmers especially the women.

The mandate of the Central poultry Development Organisation has been specifically revised, by restructuring all poultry units of this Department to focus on improved indigenous birds, which lay on an average 180-200 eggs per annum and have a vastly improved FCR ratio in terms of feed consumption and weight gain.

The Central Poultry Development Organisations have been entrusted with the responsibility of producing excellent germplasm in the form of day-old chicks and hatching eggs of these varieties like Nirbheek, Hitkari, Vanaraja, Shyama Cari, CHABRO, etc.

Besides, these Organisations are imparting training to approximately 1200-1500 farmers every year both in-situ and ex-situ. These Organisations are also playing a crucial role in analysing feed samples.

These Organisations, besides the activities stated above, scaling-up of diversification of other avian species like Ducks/Turkeys/Guinea fowl/Japanese Quail, upgradation of Training Unit into International Tropical Avian Management Institute in which private-public partnership is envisaged. Presently these Organisations are also supporting and hand-holding the Centrally sponsored Scheme, "Assistance to State Poultry Farms".

The Random Sample Poultry Performance Testing Centre (RSPPTC), Gurgaon is entrusted with the responsibility of testing the performance of layer and broiler varieties. This Centre conducts one layer test for the period from 0-72 weeks of age wherein the production parameters of the laying birds received from different participants comprising of private and public sector agencies are tested.

At the end of the test the data are analysed and the report indicating the relative performance of these entrants is published. Accordingly two broiler tests are conducted every year from 0-8 weeks of age and the results are published giving full information about growth rate, feed efficiency and margin of profit from each entrant. The report is published for the utilisation of the poultry industry in general, and the breeder participants in particular. This Centre gives valuable information relating to the different genetic stock available in the country.

A new Centrally sponsored scheme called Assistance to State Poultry Farms is being implemented during the Tenth Plan where one time assistance is provided to suitably strengthen the farms in terms of hatching, brooding and rearing of the birds with provision for feed mill and their quality monitoring and in-house disease diagnostic facilities. During the first three years of the Tenth Plan, Rs 29.75 crore has been released for assisting around 60 farms, some partially. A new scheme "Dairy/Poultry Venture Capital Fund" has been launched during the year 2004-05, wherein there is a provision to grant subsidy on interest payment. Components like establishment of poultry breeding farm with low input technology birds and also for Ducks/Turkey/Guinea Fowl/Quail/ Emu/Ostrich, establishment of feed godown, feed mill, feed analytical laboratory, marketing of poultry products (specialised transport vehicles, cool room storage facilities and retention sheds for birds, etc.), Egg grading, packing and storage for export capacity, Retail poultry dressing unit (300 birds per day), Egg/broiler carts for sale of poultry products and Central grower unit are considered for financing under this scheme.

The nodal agency for the implementation of this scheme is NABARD through nationalised commercial banks. In 2005-06, a total of 49 poultry units involving 2.17 crore was approved.

- *Sheep Development:* According to Livestock Census 2003, there are about 61.47 million sheep and 124.36 million goats in the country. About five million households in the country are engaged in the rearing of small ruminants (sheep, goats and rabbits) and other allied activities. The production of wool was 44.50 million kg during 2004-05. The expected wool production during 2005-06 stands at 50.0 million kg.

  The Central Sheep Breeding Farm, Hissar is producing acclimatized exotic/cross bred superior quality rams. The farm has supplied 1370 rams and 50 Rambouillet ewes during 2004-05 to different states. During 2005-06, the farm has supplied 854 rams and 30 crossbred ewes. The target for net production and sale of rams has been kept at 900.
- *Conservation of Threatend Breeds:* Population of some of the pure-bred small ruminants, equines, pigs and pack animals has come down considerably and such breeds have come to the category of threatened breeds in the country. The farms or the farmers unit in their respective breeding tract are to be established with cent per cent Central assistance for breeds of these animals wherein their population is less than 10,000 with active participation of State Governments and NGOs, etc.

A new Centrally Sponsored Scheme for conservation of such threatened breeds has been started during Tenth Five-Year plan with a budget outlay of Rs 1500 lakh. During 2005-06, an amount of Rs 406.92 lakh has been released for conservation of Terrasa Goat (Rs 20.00 lakh) in Andaman & Nicobar Islands, Bhutia Pony (Rs 44.39 lakh) in Arunachal Pradesh, Yak (Rs 13.00 lakh) in Himachal Pradesh, Zanskari Horse (Rs 29.96 lakh) and Double Humped Camel (Rs 29.85) in Jammu & Kashmir, Madgyal Sheep (Rs 64.50 lakh) in Maharashtra, Jamunapari Goat (Rs 59.87 lakh) in Uttar Pradesh, Gray Sindhi Horse (Rs 19.70

lakh) in Punjab, and Ghoongro pig (Rs 31.15 lakh) Bonpala Sheep (Rs 30.00 lakh) and Garole Sheep (Rs 32.25 lakh) & Black Bengal Goat (Rs 32.25 lakh) in West Bengal. There is a budget provision of Rs 400 lakh under the scheme during 2006-07.

- *Meat Production/Processing and Export:* A Centrally Sponsored Scheme "Assistance to States for modernisation/improvement of abattoirs and establishment of carcass utilisation centres and primary hide flaying units" was implemented. The objectives of the scheme were to provide wholesome and hygienic meat, gainful utilisation of animal by-products, prevention of environmental pollution and cruelty to animals. Under the scheme, financial assistance of Rs. 14.15 crore was provided for incomplete projects to States of Arunachal Pradesh, Maharashtra, Rajasthan, Andhra Pradesh, Tamil Nadu, Uttar Pradesh, Nagaland, Kerala, Tripura and Union Territory of Chandigarh during 2004-05. The scheme has now been discontinued as per Zero Based Budgeting (ZBB) decision.
- *Piggery Development:* There are more than 128 lakh pigs in the country of which approximately 14.5 per cent are graded and exotic variety. There are about 158 pig breeding farms in the country run by the State Governments/Union Territories. Exotic breeds like Large White Yorkshire, Hampshire and Landrace are maintained at these farms.

A Centrally-sponsored Scheme, namely, Assistance to States for Integrated Piggery Development, was implemented for strengthening the pig breeding farms in States during Ninth Plan. During the period, an amount of Rs 1521.87 lakh was released for strengthening the pig breeding farms in various States. The scheme was discontinued at the start of the Tenth Plan. A new scheme on piggery development viz. 'Integrated Piggery Development' has been initiated, on the basis of evaluation of the earlier scheme by NABARD, during 2006-07. Integrated piggery development scheme will be part of a new restructured Centrally sponsored macro-management scheme viz. National Project for Improvement of Poultry and Small Animals.

- *Feed and Fodder Development:* For expression of the production potentiality of our livestock, availability of nutritious feed and fodder is essential. To facilitate fodder availability, seven regional stations have been established in different agro-climatic zones for production of high yielding varieties of fodder/fodder seed and transfer of scientific fodder production technology through training of field officers, demonstration of latest fodder agronomy practices and organisation of farmer training/ fair. During 2004-05, these stations produced 199.54 MT of fodder seeds and conducted 3862 field demonstrations of new fodder varieties.

A Central fodder seed production farm located at Hessarghatta (Karnataka) produced 76 MT seeds of different varieties of fodder grasses/legumes seeds during 2005-06 and conducted 540 field demonstrations of new fodder varieties. A Central mini kit demonstration programme of fodder crops is under implementation for popularising high yielding fodder varieties on a large-scale. During 2005-06, 3.54 lakh minikits were allotted to the States for distribution to farmers free of cost. Besides, a Centrally Sponsored Fodder Development Scheme is being implemented from 2005-06 for establishment of Fodder Block making units, Grassland development with grass reserves, fodder seed production and biotechnology research. During 2005-06, financial assistance of Rs 1162.39 lakh has been provided to 15 States.

NATURAL RESOURCE MANAGEMENT

Soil and water conservation measures are one of the essential inputs for increasing agricultural output in the country. These programmes were first launched during the First Plan. From the very beginning, emphasis has been on development of technology for problem identification, enactment of appropriate legislation and constitution of policy coordination bodies.

While conceptual framework of soil and water conservation activities has been changed, concept of programmes has undergone considerable revision during successive Five-Year Plans.

The Centrally-Sponsored Scheme of Soil Conservation in the catchments of River Valley Project (RVP) was started in the Third Five-Year Plan. Subsequently, another scheme of Flood Prone Rivers (FPR) was started in the Sixth Five-Year Plan keeping in view the magnitude of floods in 1978. Both the schemes were clubbed together during the Ninth Five-Year Plan on recommendation of Expenditure Finance Committee and were subsumed under Macro Management Mode since November 2000. Under the programme for the Catchment Management of River Valley Projects and Flood Prone Rivers, 53 Catchments are covered, spread over in 27 States.

The total Catchment area is 141 million ha. with Priority Area needing urgent treatment in 28 million ha. Out of this 6.08 million have been treated till 2004-05 with an expenditure of Rs 1894 crore. During 2005-06 an area of 0.17 million ha. have been treated with an expenditure of Rs 145 crore.

A Centrally-Sponsored Scheme of Reclamation of Alkali Soil was taken up in Punjab, Haryana and Uttar Pradesh during the Seventh Five Year Plan. During the Ninth Plan, extension of the Scheme to all other States of India was approved where alkali soil problems exist as per scientific parameters.

The Scheme aims at improving physical conditions and productivity status of alkali soils for restoring optimum crop production. The major components of the scheme include, assured irrigation water on farm development works like land leveling, bunding and ploughing, agriculture, community drainage system, application of soil amendment organic manures, etc. An area of 0.66 million ha out of 3.5 million ha., of alkali land has been reclaimed till the end of 2004-05 in the country. The Scheme at present stands subsumed within the Macro Management Scheme. During 2005-06 an area of 0.03 million ha have been reclaimed with an expenditure of Rs 9.29 crore.

The Scheme of Watershed Development Project in Shifting Cultivation Areas was launched in seven North-Eastern States during the Eighth Plan from 1994-95 with 100 per cent Central assistance to the State Plan. The Scheme aims at overall development of jhum areas on watershed basis. During the Eighth Plan an amount of Rs 31.51 crore was spent by the States of north-east region and 0.67 lakh ha. area was treated through treatment packages. During the Ninth Plan, 1.57 lakh ha. have been treated with an expenditure of Rs 82 crore. The new guidelines of the scheme on the basis of new watershed to common approach has been effective from November 2000 in the revised cost norms of Rs 10,000 per ha. on net treatable area basis with additional activities and improved institutional mechanism. During the Tenth Plan, an area of 0.89 lakh ha. has been treated with an expenditure of Rs 88.32 crore up to end of 2005-06.

RAINFED FARMING

Rainfed Farming is complex, diverse and risk prone and is characterised by low levels of productivity and low input uses. The Government of India has accorded high priority to the holistic and sustainable development of rainfed areas through integrated watershed development approach. The key attributes of the watershed approach are conservation of rain water and optimisation of soil and water resources in a sustainable and cost-effective mode. Improved moisture management increases the productivity of improved seeds and fertiliser. So conservation and productivity enhancing measures become complementary.

A National Watershed Development Project for Rainfed Area (NWDPRA) was launched in 1990-91 in 25 states and two UTs. The project was continued during Tenth Five Year Plan in 28 states inclusive of three newly created States and two UTs with the purpose of increasing agriculture productivity and production in rainfed areas. The guidelines for NWDPRA have been radically restructured by retaining the technical strength and incorporating community participation. The watershed development programme is now planned, implemented, monitored and maintained by watershed communities.

- *Graph:* During Agriculture Ninth plan, an area of 27.66 lakh ha has been treated. Now the scheme of NWDPRA has been subsumed within the Macro Management of Agriculture scheme. The scheme is being continued for implementation during Tenth plan under the revised guidelines, with people's participation, covering more than 6000 micro-watersheds. It is estimated to develop an area of about 20 lakh ha during Tenth plan. In the first four years of Tenth Plan, an area of 1.59 million hectares has been developed at an expenditure of Rs. 793.82 crore.

A Watershed Development Fund (WDF) has been established at NABARD with the objective of integrated watershed development in 100 priority districts of 18 States through participatory approach. The total corpus of WDF is Rs 200 crore. Under WDF, two-thirds of amount is given for loan based project and one-third of amount is given for grant based project in the State. A number of externally aided projects are also under implementation on watershed approach, which covers an area of about 1.5 lakh hectares annually.

INTEGRATED NUTRIENT MANAGEMENT (INM)

The main objective of Integrated Nutrient Management (INM) Division is to ensure adequate availability of quality fertilisers to farmers through periodical demand assessment and timely supply, promoting integrated nutrient management, which is soil test based, judicious and balanced use of chemical fertilisers in conjunction with organic manures and bio-fertilisers, promotion of organic farming and ensuring quality control of fertilisers through implementation of Fertiliser (Control) Order, 1985.

India is the third largest producer and consumer of fertilisers in the world after China and the USA. Against 18.4 million tonnes of fertiliser nutrients (NPK) consumed during 2004-05, the nutrient consumption is estimated at 20.2 million tonnes during 2005-06. The consumption of major fertilisers namely, Urea, DAP, MOP, SSP and Complexes are estimated to be 22.2, 6.8, 2.6, 2.8 and 6.7 million tonnes respectively during 2005-06. India is, by and large, self sufficient in respect of Urea and about 90 per cent in case of DAP. The all-India average fertiliser consumption is 96.4 kg./ha of NPK nutrients through there is wide variation from state to state varying from 197 kg./ha in Punjab, 164 kg./ha in Haryana to less than 10 kg./ha in States like Arunachal Pradesh, Nagaland, Sikkim, etc. Considering the skewed pattern of fertiliser use, Government of India is promoting balanced and integrated use of fertiliser nutrients through various initiatives. As a result, NPK consumption ratio has now improved to 5.7: 2.2: 1 during 2004-05 from 7: 2.7: 1 during 2000-01.

Price of fertilisers

Presently urea is the only fertiliser which is under Statutory Price Control. To ensure adequate availability of fertilisers to farmers at reasonable rates, subsidy is provided by Government of India. Urea, the most consumed fertiliser, is subsidised under the New Urea Pricing Scheme, whereas P&K fertilisers, which are decontrolled, are covered under the Concession Scheme. The existing scheme for special freight subsidy has been continued for supplies to the North Eastern States and Jammu and Kashmir. There is no change in the prices of major fertilisers level since 12 March 2003, which are as under:

Fertilizer Product	MRP (Rs. Per MT)
Urea	4830
DAP (Indigenous)	9350
Complexes	7100-9080
MOP	4455
SSP	Price of SSP is fixed by State Governments.

Buffer Stocking of P&K Fertilisers

To ensure adequate availability of decontrolled fertilisers in remote and inaccessible areas, there is a scheme under which a buffer stock of limited quantities of DAP and MOP is being maintained at strategic locations to meet emergent requirements.

Balanced Use of Fertilisers

The Government is implementing a Centrally Sponsored Scheme 'Balanced and Integrated Use of Fertilisers' to popularise soil test based judicious application of chemical fertilisers in combination with organic manures and bio-fertilisers. Under the Scheme, financial assistance is provided for setting up/strengthening of new soil testing laboratories for advising farmers on major and micro nutrients application, training and demonstration on balanced use of fertilisers and also for setting up of production units of useful organic manure from city waste/garbage.

This scheme has since been subsumed within 'Macro Management Scheme' with effect from October 2000. State Governments are required to undertake all these activities through their Annual Work Plans.

Fertilser Quality Control

The Government ensures the quality of fertilisers through Fertiliser (Control) Order (FCO) issued under Essential Commodities Act (ECA) to regulate the price, trade, quality and distribution of fertilisers in the country. The State Governments are the executing agencies to implement the various provisions of FCO. The Order strictly prohibits the manufacture, import and sale of any fertiliser, which does not meet prescribed standards. The Central Fertiliser Quality Control and Training Institute at Faridabad and its 3 Regional centres located at Navi Mumbai, Chennai and Kalyani have been set up for inspection and analysis of imported and indigenous fertilisers, giving technical advice and providing training on quality control to state enforcement agencies and analysts. The Institute has also developed a Quick Testing Kit for on-the-spot detection of adulteration in fertilisers.

The FCO has been recently amended to make it more user friendly and ensuring effective enforcement. The important amendments include provision of repackaging of cut or torn fertiliser bags and reprocessing of damaged fertiliser due to natural calamities and provisions for commercial trials of provisional fertilisers. Accordingly, M/s National Fertiliser Limited (NFL), M/s Tata Chemicals, M/s Chambal Fertilisers Ltd., M/s Indo-Gulf Fertilisers and M/s Shri Ram Fertilisers have been permitted to manufacture neem-coated urea under Clause 20 A for commercial trials. M/s Coromandel Fertilisers Ltd. have been permitted to manufacture Bentonite Sulphur for commercial trials and M/s IFFCO has been permitted to manufacture fortified complex fertilisers. Printing of Maximum Retail Price (MRP) on all fertilisers including urea is now mandatory along with the printing of month and year of manufacture/ import of fertiliser.

Presently there are 67 laboratories in the country (including 4 Central Government Laboratories) with a total annual analysing capacity of 1.25 lakh samples. The analytical capacity and the number of samples analysed and found non standard during the last 5 years are as under:

Year	No. of Labs.	Annual Analytical capacity	No. of samples analysed	Per centage of non standard samples
2000-01	66	120315	103436	5.3
2001-02	66	120415	104188	5.7
2002-03	67	125480	109504	5.4
2003-04	67	124778	104647	5.5
2004-05	67	124730	108859	6.0

The per centage of non standard samples on all India basis is hovering around 5-6 per cent samples during the last 5 years.

However, there is a large variation from state to state as 18 per cent in M.P., 10-13 per cent in Maharashtra, HP and UP, 4-8 per cent in Bihar, Orissa, W.B., Rajasthan, Haryana, Chhattisgarh, A.P., Kerala, Karnataka and Tamil Nadu and around 1-3 per cent in Punjab and Gujarat.

- *Promotion of Bio-Fertiliser:* With a view to promote the use of bio-fertilisers as an environment-friendly and cheaper source of plant nutrients, Government of India had earlier taken up a 'National Project on Development and Use of Bio fertilisers' during the Sixth Plan and was continued till Ninth Plan. The Scheme provided for setting up National Bio-fertiliser Development Centre at Ghaziabad and its Six Regional Bio-fertiliser Development Centres (RBDCs) at Jabalpur, Nagpur, Bangalore, Bhubaneswar, Hissar and Imphal for promotion, extension, training and demonstration of bio-fertilisers. Under the Scheme, assistance for setting up of bio-fertiliser production units was also provided. Presently, there are around 125 bio-fertilisers production units in the country with an annual capacity of 18,000 tonnes of different types of bio-fertilisers and the annual production is about 10,000 tonnes. This scheme has been subsumed under a new Central Sector Scheme "National Project on Organic Farming" from October 2004. To undertake the new activities, the names of NBDC and RBDCs have been changed to National Centre of Organic Farming (NCOF) and Regional Centres of Organic Farming (RCOF).
- *National Project On Organic Framing:* The Government is implementing a new scheme 'National Project on Organic Farming' since October 2004 with an outlay of Rs 57.05 crore for production, promotion, market development of organic farming in the country during the remaining period of the Tenth Plan. The main components of the scheme include:
  - i. Putting in place a system of certification of organic produce.
  - ii. Capacity building through service providers.
  - iii. Financial support for commercial production units for production of organic inputs like:
    - a. fruits and vegetable waste compost units;
    - b. biofertilisers production; and
    - c. hatcheries for vermi culture.
  - iv. Promotion, extension and market development of Organic Farming.

PLANT PROTECTION

Plant Protection continues to play a significant role in achieving targets of crop production. The major thrust areas of plant protection are promotion of Integrated Pest Management, ensuring availability of safe and quality pesticides for sustaining crop production from the ravages of pests and diseases, streamlining the quarantine measures for accelerating the introduction of new high yielding crop varieties, besides eliminating the chances of entry of exotic pests and for human resource development including empowerment of women in plant protection skills. Following three Plan Schemes are being implemented by Plant Protection Division of the Department of Agriculture and Cooperation through Directorate of Plant Protection, Quarantine and Storage

- Strengthening And Modernisation of Pest Managment Approach in India

Components:

A. Promotion of Integrated Pest Management (IPM)

B. Locust Control and Research

C. Training in Plant Protection

D. Implementation of Insecticides Act

On the recommendation of Planning Commission the Schemes at A to D have been merged as a single scheme i.e. "Strengthening and Modernisation of Pest Management Approach in India" with a total outlay of Rs. 99.95 crore during Tenth Five year Plan (2002-2007).

E. *Promotion of integrated Pest-Management:* Keeping in view the ill effects of chemical pesticides such as development of pest resistance, pest resurgence, outbreak of secondary pests, pesticide residues in food, fodder, soil, air and water resulting in human health hazards and ecological imbalances, Government of India has adopted Integrated Pest Management (IPM) as the cardinal principle and main plank of plant protection strategy in the country. IPM is an eco-friendly approach aimed at minimum use of chemical pesticides by employing available alternative methods for pest control like cultural, mechanical and biological and use of bio-pesticides.

To enhance the production of bio-control agents/bio-pesticides, Government of India earmarked Rs 1500 lakh as grants-in-aid during the Eighth and Ninth Plans for establishment of 29 State Bio-control Laboratories of which an amount of Rs 1372 lakh has been released to States. During the Tenth Plan, an amount of Rs 470 lakh has been approved of which Rs 405 lakh has already been released so far for the said purpose. With the adoption of IPM approach, consumption of pesticides has come down from 72,133 MT (tech. grade) during 1991-92 to 40,672 MT (tech. grade) during 2004-05 and there is consistent increase in use of bio-pesticides which are environmentally safe. In some of the States, farmers have stopped using chemical pesticides and have adopted organic farming.

F. *Locust Control and Research:* The Locust Warning Organisation, Jodhpur with its 10 Locust circle offices and 1 Field Station monitors two lakh sq. km. Scheduled Desert Area in Rajasthan, Gujarat and Haryana for locust surveillance and control. During 2005-06, an area of 150 lakh hectares has been surveyed and 24 locust bulletins have been issued. Close liaison is being maintained with other locust prone countries and the FAO to keep watch over possible locust invasion. The 24th Session of FAO Commission for Controlling the Desert Locust in South West Asia was held in New Delhi from 10-14 January 2005. During the year 2005-06, 7 Indo-Pak Border Meetings were conducted with the officers of Pakistan.

G. *Training in Plant Protection:* Until 1966, there was no national training facility in India to cater to the needs of the States. To bridge this gap, the National Plant Protection Training Institute was established in 1966 at Hyderabad (Andhra Pradesh). The Institute has been recognised as a Regional Training Centre for Plant Protection by Food and Agricultural Organisation of the United Nations and also as an Advanced Centre for training in Plant Protection Technology by the World Bank. The Institute has conducted 912 courses and trained 16,747 officers/officials including 216 foreign trainees up to 2005-06.

H. *Implementation of Insecticides Act:* Pesticides are commonly used in crop protection measures for sustaining food production. These are also used for the control of vector borne diseases. As they are meant for killing insect pests and, therefore, toxic by their nature. Their misuse or abuse may result in hazards to human beings, animals and the environment. Therefore, pesticides are required to be used safely and judiciously. Keeping this in view, their import, manufacture, sale and use, etc., are being regulated under the Insecticides Act, 1968 and the Rules framed thereunder. The provisions under the Act ensure the availability of safe, effective and quality pesticides to the farmers.

The Central Government has constituted the Central Insecticides Board under Section 4 of the Insecticides Act, 1968 to advise the Central and State Governments on technical matters arising out of the administration of this Act. The Director General, Health Services is the Chairman with 29 members. The

Central Government has constituted a Registration Committee to register pesticides after examining their formulae verifying their efficacy and safety to human beings, animals and environment.

The Government of India has also set up a Central Insecticides Laboratory with the major objectives of pre and post registration verification of the properties, performance and hazards of pesticides and the proposed use claimed by the manufacturers.

To supplement the resources of the States/UTs in the analysis of pesticides, two Regional Pesticides Testing Laboratories have also been set up at Chandigarh and Kanpur with an analysis capacity of 900 samples per annum each.

Plant Quarantine Facilities in India: Plant Quarantine regulatory measures are operative through the "Destructive insects & pests Act, 1914 (Act 2 of 1914)" in the country. The purpose and intent of this Act is to prevent the introduction of any insect, fungus or other pest, which is or may be destructive to crops.

The import of agricultural commodities is presently regulated through the Plant Quarantine (Regulation of Import into India) Order, 2003 issued under DIP Act, 1914 incorporating the provisions of New Policy on Seed Development, 1988. Further, the significance of Plant Quarantine has increased in view of Globalisation and liberalisation in International trade of plants and plant material in the wake of Sanitary and Phytosanitary (SPS) Agreement under WTO.

The phytosanitary certification of agricultural commodities being exported is also undertaken through the scheme as per International Plant Protection Convention (IPPC), 1951.

Bibliography

Aaboe, Asger: *Episodes from the Early History of Mathematics*, New York, 1964.

Akhil Baruah: *Advanced Morphology of Angiosperms*, Aavishkar, Delhi, 2008.

Bag, A. K.: *Mathematics in Ancient and Medieval India*, Varanasi: Chaukhambha Orientalia, 1979.

Bahar A. Siddiqui and Samiullah Khan: *Plant Breeding Advances and in vitro Culture*, CBS, Delhi, 1997.

Banki, L.: *Bioassay of Pesticides in the Laboratory*, Akademiai Kiado, Budapest, 1978.

Boyer , J.S.: *Measuring the Water Status of Plants and Soils*, Academic Press, N.Y., 1995.

Clawson, Calvin: *The Mathematical Traveller*, New York, Plenum Press, 1994.

Collymore L.: *Fruit Production in Barbados*, Port of Spain, Trinidad and Tobago, 1996.

Currah L. and Proctor F. J.: *Onions in Tropical Regions*, Kent, Natural Resources Institute, 1990.

Daniells J.: *Illustrated Guide to the Identification of Banana varieties in the South Pacific*, Canberra, ACIAR, 1995.

David White: *The Physiology and Biochemistry of Prokaryotes*, Oxford University Press, Delhi, 2007.

Devyani Khemka: *Animal Physiology*, Dominant, Delhi, 2003.

Dharamvir Hota: *Modern Biotechnology in Plant Breeding*, Gene Tech Books, Delhi, 2007.

Diederichsen A.: *Coriander (Coriandrum sativum L.)*. Rome, International Plant Genetic Resources Institute, 1996.

Dilke, Oswald Ashton Wentworth: *Mathematics and Measurement*, Berkeley, CA: U. of Cal. Press, 1987.

Dinabandhu Sahoo: *Farming the Ocean : Seaweeds Cultivation and Utilization*, Aravali, Delhi, 2000.

Doijode S. D.: *Seed Germination in Fruits*, New Delhi, Malhotra Publishers, 1993.

Dudley, E.: *The Critical Villager: Beyond Community Participation*, London, Routledge, 1993.

Featherly H. I.: *Taxonomic Terminology of the Higher Plants*, USA, Iowa State College Press, 1954.

Ferentinos L.: *Proceeding of the Sustainable Taro Culture for the Pacific Conference*, Honolulu, HITAHR, 1993.

Fortuner, R.: *Nematode Identification and Expert System Technology*, New York, Plenum Press, 1988.

Fowler, David: *The Mathematics of Plato's Academy*, Oxford, Clarendon, 1999.

Hardy B.: *Biology and Agronomy of Forage Arachis*, Cali, International Centre for Tropical Agriculture, 1994.

Harsh Vardhan Bhaskar: *Animal Physiology*, Campus Books, Delhi, 2008.
Hartley W.: *A Checklist of Economic Plants in Australia*, Melbourne, C.S.I.R.O., 1979.
Herbert Meschkowski: *Ways of Thought of Great Mathematicians*, Holden-Day Inc., San Francisco, 1964.
Herklots G. A. C.: *Vegetables in South East Asia*, London, George Allen & Unwin Ltd., 1972.
Kim J. S.: *Sources of Resistance to Viruses of Pepper (Capsicum spp.): a Catalogue*, Taipei, AVRDC, 1994.
Kroll R.: *Cut Flowers*, Wageningen, CTA, 1995.
Savindra Singh, H.S. Sharma and Sunil Kumar De: *Geomorphology and Environment*, ACB Pub, Delhi, 2004.
Sheela Srivastava; *Genetics Embryology and Fishes*, Mangalam Pub, Delhi, 2009.
Shivanand Tolanur: *Practical Soil Science and Agricultural Chemistry*, International Book Distributing, Delhi, 2004.
Siddiqi, M.R.: *Tylenchida Parasites of Plants and Insects*, Wallingford, UK, CAB International, 1986.
Singh H. P. and Chadha K. L.: *Genetic Resources of Citrus*, New Delhi, Malhotra Publishing House, 1993.
Sinha G. C., Reddy Y. T. N. and Singh G.: *Propagation and Rootstocks in Guava*, New Delhi, Malhotra Publishers, 1993.
Sohrab Ali: *Practical Animal Breeding*, Arise Pub, Delhi, 2008.
Stover R. H. and Simmonds N. W.: *Bananas*, United Kingdom: Longman Scientific and Technical, 1991.
Subhash Ranade and Sunanda Ranade: *Concept of Ayurvedic Physiology*, Narendra Prakashan, Delhi, 2003.
Sultan Singh and I S Pawar: *Trends in Wheat Breeding*, CBS Pub, Delhi, 2006.
Tawde A. B.: *Propagation and Rootstocks of Mango*, New Delhi, Malhotra, 1993.
Thomas E.: *Fruit Production in St. Kitts and Nevis*, Port of Spain, IICA, 1996.
Urton, Gary: *The Social Life of Numbers*, Austin, University of Texas Press, 1997.
Vanderplank, J.E.: *Plant Diseases: Epidemics and Control*, New York, NY, USA, Academic Press, 1963.
Vikas Chaudhary: *Entomology and Pest Management*, Navyug, Delhi, 2008.
Whealy K.: *The Garden Seed Inventory*, Decorah, Seed Saver Publications, 1988.
White, G.F.: *Natural Hazards: Local, National, Global*, Oxford University Press, New York, 1974.
Woolfe Jennifer A.: *The Potato in the Human Diet*, Cambridge, Cambridge University Press, 1989.
Yoshida T.: *Cultivation of Citrus Genetic Resources for Evaluation of Characteristics*, Tokyo, JICA, 1996.

Index

N

P

R

S

T

U

V

W

❑❑❑